Reviews and critical articles covering the entire field of normal anatomy (cytology, histology, cyto- and histo-chemistry, electron microscopy, macroscopy, experimental morphology and embryology and comparative anatomy) are published in Advances in Anatomy, Embryology and Cell Biology. Papers dealing with anthropology and clinical morphology that aim to encourage co-operation between anatomy and related disciplines will also be accepted. Papers are normally commissioned. Original papers and communications may be submitted and will be considered for publication provided they meet the requirements of a review article and thus fit into the scope of "Advances". English language is preferred, but in exceptional cases French or German papers will be accepted.

It is a fundamental condition that submitted manuscripts have not been and will not simultaneously be submitted or published elsewhere. With the acceptance of a manuscript for publication, the publisher acquires full and exclusive copyright for all languages and countries.

Twenty-five copies of each paper are supplied free of charge.

Manuscripts should be addressed to

Prof. Dr. F. **BECK,** Department of Anatomy, University of Leicester, 6 University Road, GB-Leicester LE1 7RH

Prof. W. **HILD,** Department of Anatomy, Medical Branch, The University of Texas, Galveston, Texas 77550/USA

Prof. Dr. J. van **LIMBORGH,** Universiteit van Amsterdam, Anatomisch-Embryologisch Laboratorium, Mauritskade 61, Amsterdam-O/Holland

Prof. Dr. R. **ORTMANN,** Anatomisches Institut der Universität, Lindenburg, D-5000 Köln-Lindenthal

Prof. J.E. **PAULY**, Department of Anatomy, University of Arkansas for Medical Sciences, Little Rock, Arkansas 72205/USA

Prof. Dr. T.H. **SCHIEBLER,** Anatomisches Institut der Universität, Koellikerstraße 6, D-8700 Würzburg

Advances in Anatomy
Embryology and Cell Biology

Vol. 78

Editors
F. Beck, Leicester W. Hild, Galveston
J. van Limborgh, Amsterdam R. Ortmann, Köln
J.E. Pauly, Little Rock T.H. Schiebler, Würzburg

Gerd Grün

The Development
of the Vertebrate Retina:
A Comparative Survey

With 15 Figures

Springer-Verlag
Berlin Heidelberg New York 1982

Dr. Gerd Grün
Spezielle Zoologie
Ruhr-Universität Bochum
D-4630 Bochum 1
FRG

Present Address:
Steinkuhlstr. 10
D-4630 Bochum 1
FRG

ISBN-13:978-3-540-11770-4   e-ISBN-13:978-3-642-68719-8
DOI: 10.1007/978-3-642-68719-8

Library of Congress Cataloging in Publication Data
Grün, Gerd, 1945 – The development of the vertebrate retina.
(Advances in anatomy, embryology, and cell biology; v. 78)
Bibliography: p. Includes index. 1. Retina. 2. Embryology-Vertebrates.
I. Title. II. Series. [DNLM: 1. Retina. 2. Cell differentiation. W1 AD433K
v. 78 / WW 270 G926d]
QL 801.E67 vol. 78 [QL949] 574.4s 82-10440
ISBN-13:978-3-540-11770-4 (U.S.) [596'.0332] 82-10440

Composition: Schreibsatz Service Weihrauch, Würzburg

2121/3321-543210

Contents

Acknowledgments

My thanks are due to all the researchers troughout the world who sent me offprints of their papers, to Drs. Schwemer, Nilsson, and Höglund for permission to read their unpublished paper, and above all to the staff of the Embryological Library of the Hubrecht Laboratory at Utrecht, Netherlands, without whose kind help it would not have been possible to write this survey.

1 Introduction

The mature vertebrate retina is a highly complicated array of several kinds of cells, capable of receiving light impulses, transforming them into neuronal membrane currents, and transmitting these in a meaningful way to central processing. Before it starts to develop, it is a small sheet of unconspicuous cells, which do not differ from other cells of the central nervous system. The chain of events which lead to the transformation from this stage into that of highly specialized cells ready to fulfil a specific task, is usually called "differentiation." Originally, this word indicated firstly the process of divergence from other cells which were previously alike, and secondly, the change from an earlier stage of the same cell. It has become widespread practice to imply by the word "differentiation" also the acquisition of specific properties and capacities which are characteristic of a mature, i.e., specifically active, cell.

Every cell is active at any stage of development, but certain activities are shared by most cells (e.g., the activities of preparing and accomplishing proliferation, that of initiating development, that of maintaining a certain level of metabolism), while there are others which are shared by only a small number of — originally related — cells. In most cases these latter activities are acquired by the final steps of cellular development, the terminal "differentiation." In the context of the present paper, the word "function" will refer to this latter type of specific activity. Together with final structures it characterizes a certain type of cells and discerns them from all others.

What has been said here about the cell applies to a tissue too. But is tissue differentiation more than, or different from, the sum of the differentiation of its cells? Can all events in the cells, when summed up, explain the specific function and its acquisition in the tissue? This question will be studied in the vertebrate retina, where tissue function is different from the function of each of its cells.

Some cells of the retina transform certain properties of certain electromagnetic waves into membrane currents; others transmit currents by means of specific transmitter substances which they are able to synthesize as well. The retina as a whole is capable of transmitting a continuous representation of the light conditions of the surrounding space to the brain, and it does so by means of coordinated patterns of substances and membrane potentials, characteristic of the retina only, but not of any of its single cells.

On the other hand, differentiation occurs in single cells. It is the aim of the present review to show that the summation of single cell events leads to the specific functioning of the retina as a whole.

By their nature and their results, the processes of retina development can be arranged in different phases, which also reflect the order by which developmental processes in the retina as well as in other neural tissues are observed. These phases are:

1. Proliferation and migration
2. Forming of final cell and tissue shape; "layering"
3. General differentiation, i.e., differentiation of structures and processes on the cellular and tissue level, which are not characteristic of specific retinic or neuronal activity but reflect the maturation of common cellular properties

4. Differentiation of specific neuronal and photoreceptoral structures
5. Differentiation of specific retinic or neuronal substances and chemical processes
6. Differentiation of specific retinal or neuronal activities

Though retina development has been studied in a number of animals (ca. 50 species), it is only in a very restricted number that a substantial body of information has been collected for all these phases, and therefore only in a restricted number of species does a comparative survey seem possible. Here the development of retinal cells is looked at in man and those five species which represent different classes of vertebrates and also on which the highest number of pertinent studies has been carried out: the rat (*Rattus norvegicus*), the mouse (*Mus musculus*), the chick (*Gallus gallus domesticus*), the clawed "toad" (*Xenopus laevis*), and the cichlid fish (*Tilapia leucosticta*).

From the mass of information obtained in these species, this review aims, by means of comparison, to trace temporal and other correlations, sequences, regularities, and differences in the events which lead to one and the same result: the specifically active retina.

In order to gain a more detailed picture of the way this result is attained, the events of phases 4, 5, and 6 will then be studied more comprehensively. Based on nearly the entire amount of available literature, the maturational processes of those structures and functions which are highly specific for the retina are traced back. Thus, items studied in more detail are: the receptor cell inner segment, the receptor cell outer segment, the receptor cell terminal and the inner plexiform layer (phase 4); retina neurochemistry (phase 5); and the phenomenon of retinomotoric behavior, certain aspects of ganglion cell physiology, and the electroretinogram (phase 6).

The formation of tractus opticus junctions in the optic centers and all the questions concerning retinotectal connections and specification will not be included. They have been competently reviewed several times (Jacobson 1976b; Edds et al. 1979; Rager 1979). Furthermore, questions of the ability to "see," of increasing acuity [see, e.g., Ordy et al. 1965; Easter et al. 1977; a summarizing work concerning the human eye is that of Dobson and Teller (1978)], and of correlation to biological topics (e.g., Baburina 1972) are not included. All these questions involve nonretinal factors, and therefore maturational steps cannot be judged adequately. Within the retina, mainly neurons and receptor cells have been taken into consideration. The glial Müller cells and the pigmented epithelium are only mentioned peripherally.

2 Temporal Patterns and Correlations

The mature retina, the result of particular differentiational events, is a unit, i.e., a spatial unit of clear-cut, determined connections, patterns, and relationships, as well as a functional unit. The spatial order, which is attained during the process of development, has its meaning only in the function of the entire unit, a function which does not exist without the achievement of the spatial order. Structures differentiate to fulfil a task within a functional unit, which is not yet present at the time of their formation, and which necessarily relies on other structures to be formed at different sites within the unit. This implies that the function of the whole unit is present as soon as the last single structure has been formed, thus completing a spatial arrangement of particular functions.

This, together with the need of a coordination between the various structures arising at different locations in the retina, requires a temporal order as well. This temporal order — basic and prior to the spatial one — will be treated in the following chapters.

2.1 Differentiation of the Retina in *Tilapia leucosticta*

The cichlid *Tilapia leucosticta* is the teleostean species in which most research on cytological differentiation of the retina has been done. In the following, published as well as unpublished results of the present author are summarized; therefore no references will be given. The differentiation up to the attainment of visual function in the retina takes only few days and is highly dependent on temperature. Batches differ in time-stage relations, which renders a separation into temporal units smaller than days meaningless.

In this and the following chapters day or days will be abbreviated to d, days posthatch or after birth will be denoted by d..pp, W = week(s), and M = month(s).

2.1.1 Proliferation

In the course of d4, the number of mitoses decreases abruptly. This reduction continues during the following days, though at a lowered rate. Mitoses are found up to d9, but after d4 they are confined to the more peripheral regions.

2.1.2 Cell and Tissue Shape; Layering

The layering of ganglion cells begins on d4 and is followed by an inner plexiform layer on d5. At the end of d5, the nuclei of amacrine and bipolar cells are different, and a short time later, horizontal cells can be identified.

2.1.3 General Differentiation

On d4, the fusiform cells of the retina are in an "undifferentiated" state, as described for numerous other types of neuroblasts (see Fig. 1). They show numerous ribosomes and some mitochondria. One day later, the number of ribosomes in ganglion cells is reduced while endoplasmic reticulum and mitochondria increase and a Golgi apparatus appears. The amacrine cells of that stage are very similar to ganglion cells, but obviously they contain more ribosomes. Yet the ribosomes in the bipolar cells are more abundant. In addition ER and mitochondria begin to appear in these two types of neuroblasts, and similar tendencies are found in receptor cells. By d7, the amount of ribosomes and mitochondria decreases in bipolar cells.

On the 5th and 6th days, several types of junctions, with or without cleft densities, are found to interconnect processes of the intermediate neurons and the ganglion cells. Acid phosphatase can be detected in the inner plexiform layer on d8. (This enzyme is especially connected with the stages of final neuronal maturation, cf. Wender 1972.)

On day later, alkaline phosphatase, an enzyme which may be regarded as the marker of an enhanced neuronal metabolism (Tewari and Tyagi 1968) is present in the inner plexiform layer.

3

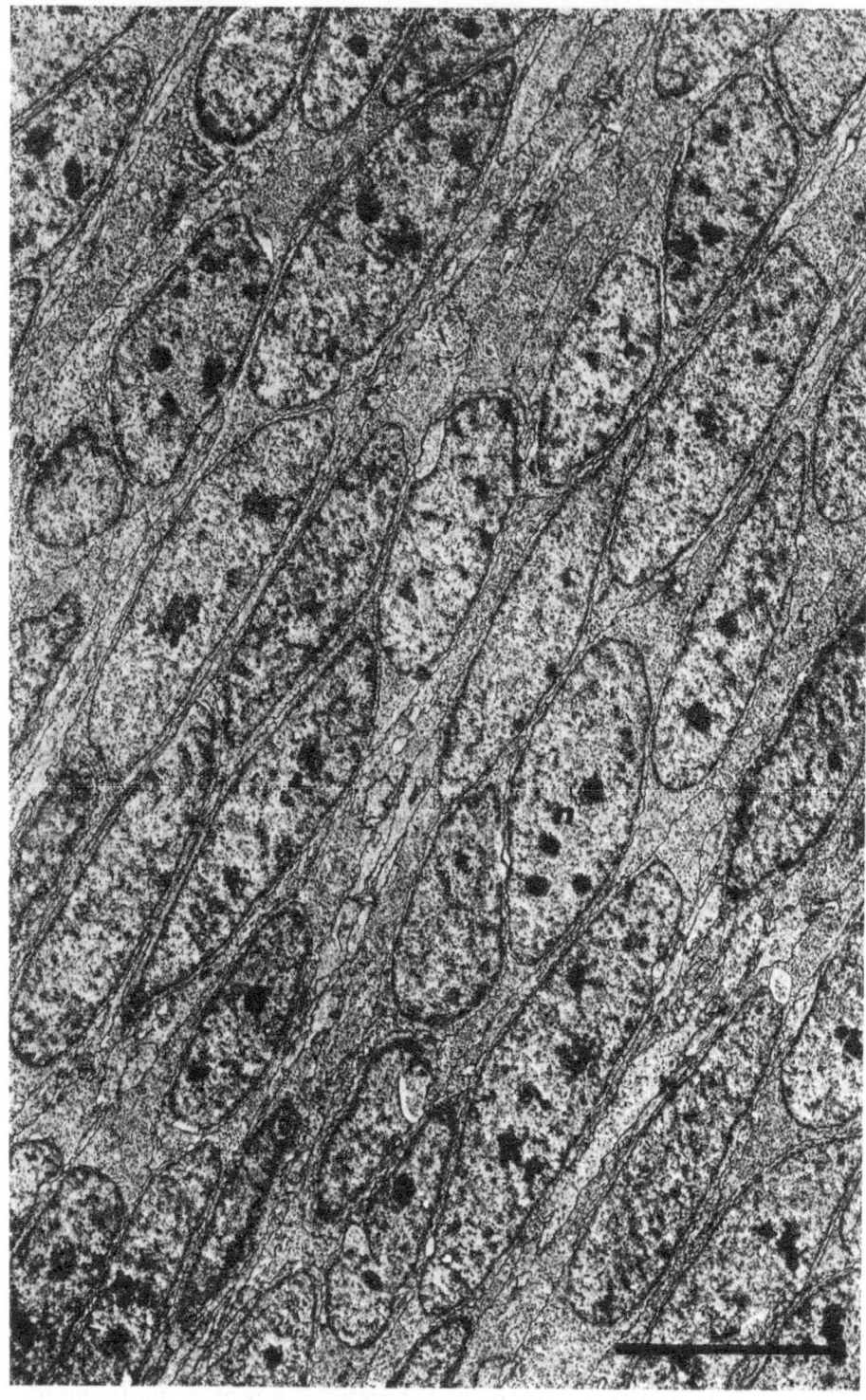

Fig. 1. Undifferentiated neuroretina of an 8-day-old chicken embryo. Cells are fusiform and contain mainly ribosomes. *n*, nucleus. Bar = 5 μm. (Meller 1968)

2.1.4 Specific Structural Differentiation

The first ganglion cell neurites grow out on d4. While the inner plexiform layer is found to be composed of short processes of ganglion and amacrine cells on d5, the bipolar cells form axonal growth cones with numerous ribosomes, neurofilaments, and single ER membranes not earlier than at the end of this day. Neurotubuli are found in the inner plexiform layer on d6, and, more neurites and synaptic bands on d7. Mature conventional synapses are formed by the end of this day, connecting amacrine to amacrine cells. The axon terminals do not contain more than 20 synaptic vesicles per ultrathin section.

On d8, increasing numbers of junctions aquire pre- and postsynaptic dense projections and a mature aspect. The first ribbon synapses are found. Amacrine cells now from junctions with ganglion and bipolar cells. The number of synaptic vesicles in amacrine cell terminals amounts to nearly 50 per ultrathin section.

In the second half of d6, horizontal as well as bipolar cells form dendrites which penetrate the receptor terminal, thus forming triadic configurations. In the receptor terminal, precursors of the synaptic lamellae can be found for a very short period, and the first synaptic vesicles appear. One day later the synaptic lamellae in the receptor terminals have grown to attain a length of more than 1000 nm, and the number of synaptic vesicles has increased to that of adult level, while ribosomes are reduced. By d9, the distribution and arrangement of synaptic lamellae in receptor cell terminals is adultlike.

In the receptor cell inner segments, an accumulation of mitochondria is noted on days 5 and 6. Then, in the inner segments of the double cones, subsurface cisterns are formed by the fusion of vesicles. On d7, an ellipsoid has formed besides numerous ER membranes and Golgi vesicles.

A ciliary stalk projects from the inner segment in the second half of d6, and – synchronous to an enlargement of its scleral part –, the first disk precursors are seen to invaginate from its membrane. As soon as d7 the outer segments of both rods and cones comprise numerous disks, and phagosomes are already found by d8, when outer segments are still growing.

2.1.5 Neurochemical Differentiation

On d7, the presence of catecholamines in the vesicles of amacrine cell terminals can be demonstrated by ultracytochemical methods. For the first time, and from now on regularly, adenosine triphosphatase is localized cytochemically in synaptic vesicles and at the terminal membranes of the inner plexiform layer, as well as in mitochondria and at subsurface cisterns of the inner segments. One day later, visual pigment can be detected microspectrophotometrically. The distribution and the amount of ATPase in the inner plexiform layer, and in the inner and outer segments is adultlike.

2.1.6 Differentiation of Specific Activity

With the beginning of d9, pigment migration can be found.

2.2 Differentiation of the Retina in *Xenopus laevis*

Although, according to Nieuwkoop and Faber (1967), the generally adopted system of staging for this anuran amphibian had not been designed to meet the needs of a staging of the retina development, it will be applied here because it gives a widely known framework for the subdivision of this short developmental phase.

2.2.1 Proliferation

At stage 28 (d2) the retina consists of replicating cells, but 2 h later, at st. 29, the DNA synthesis has ceased in the ganglion cells (Jacobson 1968b). Shortly before metamorphosis, at st. 54, mitotic activity decreases in the dorsal half of the retina and increases in the ventral half. This change is due to different reactivities to thyroxin (Beach and Jacobson 1979a,b).

2.2.2 Cell and Tissue Shape; Layering

Up to st. 32 (d2) all cells of the retina resemble each other (Fisher and Jacobson 1970). A distinctly visible inner and an outer plexiform layer appear at st. 37 (d3). Both layers increase in thickness at least until st. 45 (d5) (Chung et al. 1975; Fisher 1976). With st. 48 (d8) the retina is adultlike in appearance (Fisher and Jacobson 1970). Cones and rods are discernible in light microscopic preparations from st. 41 (d4) on (Saxén 1954).

2.2.3 General Differentiation

Up to st. 32 (d2) retina cells are embryonic in aspect and show free ribosomes. There is no sign of differentiation (Dixon and Cronly-Dillon 1972; Chung et al. 1975), but the amount of helical polysomes, subsurface cisterns, and junctions has increased from st. 28 (d2) on (Grillo and Rosenbluth 1972), and the endoplasmic reticulum, Golgi apparatus, microtubuli, neurofibrils, and Nissl bodies now make their appearance in the ganglion cells (Fisher and Jacobson 1970), as well as in the cells of the outer retina (Kinney and Fisher 1978b).

The junctions observed in the retina after st. 26 (d2) are zonulae adhaerentes, macula adhaerentia, and gap junctions (Dixon and Cronly-Dillon 1972), the latter being present between the ventricular cells of the future pigmented epithelium. Up to st. 32 (d2) they are more frequent in the dorsal half of the retina, thereafter in the ventral half (Hayes 1976). They are also said to connect future receptor cells, while Kinney and Fisher (1978b) attribute this role to the zonulae adhaerentes. Stage 38 (d3) is the last where gap junctions are found within the neuronal retina, but at stage 41 (d4) they have reappeared to connect the inner segments of neighboring receptor cells (Hayes 1976).

A specific 48 000-dalton protein is found in the retina at st. 28; at st. 29, 2 h later, it is confined to the ventral half of the eye cup and has disappeared by st. 32 (d2) (Cafferata et al. 1979). As soon as st. 31 (d2), a general increase of retina protein begins, resulting in a 30-fold rise in the amount up to st. 63 (midway metamorphosis) (Ma and Grant 1978). Up to st. 47 (d6) this is paralleled by the incorporation of

³H-glucosamine into the region of the future plexiform layers, where it will serve as a component in the glycocalyx of the large number of cellular processes to come (Hollyfield et al. 1975). At st. 37 (d3) the lectin concanavalin A is suddenly bound intensely to the inner and the outer plexiform layer and to ganglion cell neurites (Ulshafer and Clavert 1979).

2.2.4 Specific Structural Differentiation

Beginning with st. 28 (d2) the ganglion cells of the centralmost region produce neurites which contain neurofibrils and microtubuli and are already fasciculated (Cima and Grant 1980). They grow up to the optic chiasma until st. 35 (Chung et al. 1975). At this stage (d2), 151–191 fibers have been counted; at st. 38 (d3) the tractus opticus is made up of less than 1000 fibers. They are 0.1 – 0.5 μm thick and not yet myelinated, but already fasciculated. Diameters are found to differ even more at st. 46 (d5), when there are 1500 – 2000 neurites in the tract (Wilson 1971). Myelination begins at st. 49 (d12) (Gaze and Peters 1961), but at st. 52 (d21) the number of myelinated fibers still amounts to less than 100, while 10000 nonmyelinated fibers are found (Wilson 1971). The thickness of the tractus opticus increases even after metamorphosis (Chung et al. 1975), presumably because of myelination.

Between st. 35 and st. 40 (d3) the newly formed neurites establish topographically organized connections with the tectum opticum (Jacobson 1968a). The position of the ganglion cells with respect to the tectum, however, is specified as early as st. 31 (d2) and before (Jacobson 1968a), or st. 32 (d2) (Dixon and Cronly-Dillon 1972; Grillo and Rosenbluth 1972), that is, before the first neurites extrude. Specification can even be shifted to st. 24/25 (d2), when the ionophore X 537 A is administered to the embryos (Jacobson 1976a).

The first synapses in the inner plexiform layer are found at st. 40 (d3), among those are also ribbon synapses of bipolar cells. A slow but constant increase of these starts now and lasts up to st. 47 (d6), when the first series seems to be completed (Fisher 1976). Tucker and Hollyfield (1977), however, report no change in conventional and ribbon synapse density after st. 43 (d4); serial synapse density increases to adult level at the end of metamorphosis. Initial contacts between the receptor terminal and horizontal cells are found at st. 37/38 (d3), consisting mostly of one postsynaptic process; synaptic vesicles are still sparse and form a cluster around the short synaptic ribbon. In st. 39, 4 h later, horizontal cell dendrites grow to form long digitiform processes, which invaginate the terminals and will be postsynaptic to single ribbons at st. 40. Then they enormously increase in length, and arciform densities appear. The number of bipolar cell/receptor terminal contacts is very low during these stages, and even at st. 42 (d4) they are only superficial, forming "basal junctions" (Chen and Witkovsky 1978).

In the inner segment, from st. 41 (d4) on, a myoid and an ellipsoid are well developed (Saxén 1954). Beginning with st. 37/38 all receptor cell inner segments contain an oil droplet (Kinney and Fisher 1978b), and no sooner than st. 43 (d4) rod inner segments have lost it (Witkovsky et al. 1976).

The first sign of an outer segment is found at st. 35/36 (d3), when a ciliary stalk begins to grow out of one centriole (Kinney and Fisher 1978b). At st. 37/38 (d3) first disks are reported to occur in some receptor cells, the number of which soon increases (Witkovsky et al. 1976). For this stage Kinney and Fisher (1978b) describe outer segments as being all conical in shape. According to them, rod outer segments display discrete bands of label after injection of ^{3}H-leucine, while in the cones the label is diffuse. The tapering shape of the cone outer segment becomes evident at st. 42 (d4) (Chen and Witkovsky 1978). The length of rod outer segments is 33 μm then, and the fraction of the retina occupied by outer segments is 6.6% (Witkovsky et al. 1976). The adult rod outer segments length (50 μm) is attained at st. 54 (d26) (Kinney and Fisher 1978b).

The first phagosomes are seen within the pigmented epithelium at st. 46 (d5) (Kinney and Fisher 1978b; Hollyfield and Rayborn 1979). At st. 47 (26 h later) disks are shed at a rate of 0.21 μm/d, and at st. 54, the rate of membrane addition is 1.6, that of disk shedding and phagocytosis is 1.2 μm/d (Kinney and Fisher 1978b).

2.2.5 Neurochemical Differentiation

Only a few hours after the appearance of the first disks, traces of visual pigment (0.02 pmol = 6 × 10^9 molecules per eye) could be extracted at st. 39 (d3). Up to st. 58 (d44) there is a steady increase (2.4 × 10^{14} molecules/eye) which runs parallel to the number of newly formed outer segments (fraction of retina occupied by rods then: 40%), but also to the outer segment length (then: 48 μm) (Witkovsky et al. 1976; Bridges et al. 1977). While the latter authors found rhodopsin only at st. 39 and after st. 58, a visual pigment composition of 59% porphyropsin and 41% rhodopsin was reported for st. 49 (d12) by Crescitelli (1973).

Synthesis of acetylcholine has been shown at st. 35/36 (d3) for the first time. Within the following 48 h (st. 45) it increases 30-fold, while the rate of protein synthesis declines. After st. 52 (d21) up to metamorphosis, the acetylcholine content keeps pace with the amount of protein; from that fact an increase in ACh content can be deduced for stages 56–63 (Ma and Grant 1978).

Gamma aminobutyric acid (GABA) is selectively taken up at st. 31 or 32 (d2). At st. 33/34, future horizontal cells and amacrine cells are labeled by ^{3}H-GABA. At st. 35/36 (d3) the amount is 19 pmol/mg protein (9 months postmetamorphic: 127 pmol). GABA release can be elicited at st. 37/38 (d3) when a period of rapid GABA increase starts, which stops at st. 42 (d4). At st. 49 (d12) 200 pmol/mg protein are found, at st. 63, 110 pmol; this is followed by a rise, during the last metamorphic stages, of up to 180 pmol (Ma and Grant 1978; Hollyfield et al. 1979).

At st. 38, the amount of 5-hydroxytryptophan decarboxylase and other enzymes of the serotonin metabolism is found to be high, and increases even more at st. 42 (d4) (Baker and Quay 1969), i.e., at a time when rapid GABA increase stops.

2.2.6 Differentiation of Specific Activity

As a first component of the electroretinogram, PIII (or a-wave) is consistently found at st. 39 (d3), simultaneously with the appearance of the outer segments and visual pigment. Ten hours (one stage) later, a b-wave appears and increases rapidly to become a prominent component of the immature ERG at st. 41 (d3) (Witkovsky et al. 1976).

At st. 43 and 44 (d4) the first visually evoked potentials are found in the tectum opticum and a rapid increase of sensitivity to visual stimuli begins, paralleled by the development of event and dimming units as well as T-and H-type ganglion cells, and followed (st. 45, d5) by strong on and off responses to environmental illumination. While there is a reduction in the extent of the receptive fields in the retina's most central region (st. 47, d6), and no further increase in sensitivity to visual stimuli (st. 48, d8), a further type of response unit can be identified at st. 49 (d12) which is similar to the adult dimming unit, but shows a response to graded lightness as well (Chung et al. 1975). At this stage, the visually guided behavior, too, is found to begin (Gaze and Peters 1961).

2.3 Differentiation of the Retina in the Chick, *Gallus gallus domesticus*

Some of the studies on retina development in the chick apply to the staging by Hamburger and Hamilton (1951), but this is usually not the case. The latter way will be adopted here because a comparison is thus facilitated and the Hamburger-Hamilton stages are usually used for the earliest stages only. However, it must be said, by orientation to this staging system, most of the discrepancies in relating developmental events to a certain stage would definitely have been avoided, discrepancies which arise through different rearing conditions or through breeding of different strains.

2.3.1 Proliferation

Most ganglion cells enter the last cell cycle on d3 (Kahn 1974), and some of them complete the last mitosis at this date, as most of them will do within the following 4 days (Kahn 1973; Fujisawa et al. 1976a). On d3 amacrine, horizontal, and receptor cells enter their last cell cycle as well, and some of them complete it by d5. The first bipolar cells do so no sooner than d6 (Kahn 1974).

There are great non random variations in the developmental age of receptor cells, and the duration of their S-phases differs (Morris 1973, 1975). By d7 they have lost their ability to synthesize DNA (Mishima and Fujita 1978). One day later, mitotic activity ceases, in amacrine as well as in horizontal cells (Fujita and Horii 1963; Fujisawa et al. 1976a). Overall mitotic frequency decreases by d9 (Meller 1968).

2.3.2 Cell and Tissue Shape; Layering

For early morphogenetic events see Schook (1978). On d2 the cells of the optic cup are elongated, and the future retina is characterized by numerous intercellular lacunae (Hunt 1961). Degenerating cells which are in the course of being phagocyted are found on d3 (Garcia-Porrero and Ojeda 1979). On d4 the pigment cells are discernible by their cuboidal shape, differing from the retina cells (Meller 1968), which are arranged in fascicles and columns (Meller and Tetzlaff 1976).

The first ganglion cells are recognized in histological sections on d3, when a nerve fiber layer is produced (Hughes and LaVelle 1974, 1975). On d5, the retina is 100 μm thick, a proliferating tissue showing a pigmented epithelium, an outer limiting membrane, a proliferating layer, a layer of outer neuroblasts, ganglion cells, a nerve-fiber layer, and an inner limiting membrane (O'Rahilly and Meyer 1959; Cooper and Meyer 1968). By d6, the number of ganglion cells rises sharply and they are spread over nearly the entire surface of the retina (Coulombre 1955). Six types of ganglion cells are found on day 6 (Nishimura 1980).

An inner plexiform layer appears on d6 (Shen et al. 1956) or d7 and is mainly composed of dendrites, joined by punctate junctions (Hughes and LaVelle 1974, 1975). On d8, horizontal processes of ganglion and amacrine cells have appeared (Coulombre 1955). On d9, the outer plexiform layer arises (Rebollo 1963; Hughes and LaVelle 1974, 1975), and by d10 all retinal layers are well identifiable (Leplat 1914). From d14 on, two types of bipolar cells can be recognized by their difference in cytoplasmic organelles and dendritic processes (Yew and Meyer 1975).

Receptor cells are first reported for d6 (Fujita and Horii 1963). For the next day, a sensible phase for the formation of double cones can be inferred from the possibly artificial production of triple cones (Morris 1974), while receptor cell inner segments are not conspicuous before d9, when they surpass the outer limiting membrane (Hughes and LaVelle 1974), and, under the scanning electron microscope, are seen on the outer surface of the neural retina (Olson 1979). Rods and cones can be identified, light microscopically, by d10 (Weysse and Burgess 1906; O'Rahilly and Meyer 1955).

2.3.3 General Differentiation

On d2, within the retina cells, rough endoplasmatic reticulum and ribosomes are found, the latter often arranged in rosettes (Hunt 1961), and on d5, they are still typical matrix cells with many ribosomes (Mishima and Fujita 1978; see Fig. 1). Ganglion cells contain much RNA on d4 (Rickenbacher 1952) and, 1 day later, show signs of intracellular differentiation, i.e., enhanced appearance of ER and indentation of the nucleus (Mishima and Fujita 1978). On day 10, they display more RNA than the bipolar cells (Rickenbacher 1952), which then produce ER and mitochondria (Yew and Meyer 1975) but still show an undifferentiated character and many free ribosomes on d13 (Meller 1964).

Future receptor cell inner segments are characterized by centrioles, forming diplosomes from d5 on (Olson 1973), but still on d8 they appear undifferentiated, with many free ribosomes and a few mitochondria (Govardovsky and Kharkeevich 1966) and are rich in RNA at least up to d10 (Rickenbacher 1952). In the Müller cells, which are interconnected by desmosomes, mitochondria and microvilli are produced in the apical parts on d9. On d15, they contain ribosomes, filaments, and vacuoles in their processes, which are wrapped round neuronal processes in the inner plexiform layer. They have reached the outer plexiform layer by d17, when they show a growing tendency to isolate receptor terminals and neuronal processes. On d19, this insulation seems to be completed; the large vesicles and filaments always increase in number (Meller and Glees 1965).

On d2, the basal parts of the retina cells are interconnected by terminal bars or desmosomes (Hunt 1961), for d3, all cells are reported to be provided with zonulae adhaerentes, while the cells of the pigmented epithelium are connected with the retina by gap junctions. On d5, these are found at further locations within the retina (Fujisawa et al. 1976a) and interconnect future receptor cells as well (Hayes 1977). A maximum number of gap junctions is characteristic of d6; some of them are associated to zonulae adhaerentes (Fujisawa et al. 1976a); the receptor cells, however, are reported to be joined by desmosomes (Meller 1964), as will still be the case on d9 (Meller and Breipohl 1965) and d10, when they are also found between receptor and pigment cells (Govardovsky and Kharkeevich 1965). The gap junctions on d7 also connect receptor and pigment cells (Hayes 1977), but with the beginning of d8, their number is reduced within the whole retina (Meller 1968) and 1 day later, gap junctions are only found near to the outer limiting membrane (Fujisawa et al. 1976a; Hayes 1977). For day 21 (= hatch) they are reported to connect double cones, while the receptor cells are connected with Müller cells by junctions of the zonula adhaerens type (Fujisawa et al. 1976b). In the inner plexiform layer, macula adhaerentia diminuta of 0.1 μm length and 15—20 nm width are observed on d10 (Sheffield and Fishman 1970); 1 day later, in the outer plexiform layer, only punctate junctions are found (McLaughlin 1976a).

The histochemical reaction to acid phosphatase is positive in all retinal cells from d3 on, but weakens up to the hatching. From d9 on, it is localized within three strata, parallel to the layering, but in ganglion cells acid phosphatase is demonstrated only by d21 (Stefanelli et al. 1961a). Up to d19, more acid than alkaline phosphatase will be found (Lindeman 1949). Alkaline phosphatase is found in ganglion cells on d17; 1 day later it is also present in amacrine cells, and on d19 in bipolar cells (Rebollo 1955).

The activity of the pentose-phosphate shunt is highest on d7, but declines up to hatching and increases again thereafter (Masterson et al. 1978). Succinate dehydrogenase can be detected from d12 on (Kojima et al. 1957, cited after Coulombre 1961) and has attained its final distribution on d19, along with lipase and phosphatases (Macri and Rebollo 1965).

In electrophoresis of retina proteins, most anodic fractions disappear by d2, are found again 1 day later, and have disappeared again by d8 (Mikhailov and Gorgolyuk 1976). Protein and glycoproteins are transported in the ganglion cell neurites to the tectum opticum at the rate of 60 mm/d on d10, but the rate of rapid flow will increase up to 95 mm/d on d13 and to 110 mm/d on d18 (Sjöstrand et al. 1973; Marchisio et al. 1973).

From the presence of RNA and Fast Green positivity, an enhanced protein synthesis is deduced for receptor cell inner segments by d11 (Yew 1976), and the first peak of protein content in the whole retina is reported for the next day, followed by peaks on d16 and d18 (Kataoka 1955). The peak on d16 is followed by an increase in protein concentration, which stops by d18 (Suzuki et al. 1977). With the beginning of d17, there is a decrease in free amino acid content, followed by an increase about the day of hatching (d20 to d1pp). At 1Mpp, the content of free amino acids is adult-like (Pasantes-Morales et al. 1973). Between d7 and d13, qualitative and quantitative changes in glycoprotein content and composition are found (Mintz and Glaser 1978).

An increase in retinal thickness and surface on d5 (see above) is paralleled by an increase in chondroitine-4-sulphate and, to a lesser degree, chondroitine-6-sulphate on cell surfaces, indicating the growth of the retina, which is due to a growth in cell number (Morris et al. 1977).

The ability of the pigmented epithelium cells to differentiate into retina cells is lost with d3 (Orts-Llorca and Genis-Galvez 1960) or can be extended to d5. Thereafter, a part of the retina must be present in order to induce the regeneration of the retina from the pigment epithelium (Coulombre and Coulombre 1965). The ability of retina cells of transdifferentiating into pigment or lentoid cells is not restricted to the youngest retina (d4: Okada et al. 1979; d9: Clayton et al. 1977; d14: Peck 1964), but extends up to d17, when no more lentoid cells, and to d21, when no more pigment cells, are formed out of cultivated retina cells (de Pomerai and Clayton 1978).

As early as d2, a critical phase in the fixation of retinal polarity is noted (Goldberg 1976). Treatment of the retina with 6-hydroxydopamine on d3 leads to damaged receptor inner and lacking outer segments on d19; it is concluded that the drug interferes with trophic functions of the catecholamine system (Yew et al. 1974).

2.3.4 Specific Structural Differentiation

The first fibers of the tractus opticus (first ganglion cell neurites) are reported for d3. Later on this day, they are fasciculated and reach the optic stalk (Goldberg and Coulombre 1972). The first neurites reach the diencephalon by d4, and later on, at the stage of 48 somites, they reach the site of optic chiasma formation (Rogers 1957). The growth rate of the tractus opticus on d6 is 1 mm/d; the velocity of axonal transport has been measured 92 mm/d (Crossland et al. 1974). The number of neurites increases exponentially up to d8 (Rager and Rager 1978). The first neurites are expected to reach the optic tectum (Hughes and LaVelle 1975; Rager 1976) at its anteroventral part then, spreading over the tectal surface during the following 6 days (Goldberg 1974). First synapses in the tectum appear on d11 only (Rager 1976). On d7, most ganglion cell bodies and neurites, but no dendrites, may be stained by silver techniques (Shen et al. 1956).

The total number of ganglion cell neurites has amounted to 4 million on d11 and starts to decrease, a process which is finished by d18 (Rager and Rager 1978). When the connection with the tectum is interrupted, the number of ganglion cells is reduced to about 42% on d12 (McLoon and Hughes 1978). On d14, thicker axons appear in the tractus (Rager 1976), and 1 day later, there are some myelinated fibers of 0.6–1.0 μm diameter in the tractus which have become more numerous by d17 (Blozovski 1971a,b). On d18, 3% of the tractus opticus fibers are myelinated (Rager 1976) and the number of ganglion cell neurites has been reduced to 2.4 million (Rager and Rager 1978). The greater part of these is 1.0–1.5 μm thick (Blozovski 1971b). At 3Mpp, 95% of the tractus opticus fibers are myelinated and the diameter of the tractus has increased to 2.3 mm (on d8: 0.15 mm), mainly by the enhanced formation of fibers, but also by the appearance of thicker single fibers (Rager 1976).

On d8, horizontal processes of ganglion and amacrine cells constitute the inner plexiform layer which appears at that stage (Coulombre 1955). They form one, and later on, two synaptic bands (Shen et al. 1956). During the preceding 2 days certain retinal neurons are able to make synapses onto muscle cells, which allows one to conclude that synapse specificity is attained only after this phase (Ruffolo et al. 1978). On d9, ganglion cell dendrites branch within the inner plexiform layer and make synapses with bipolar cell terminals within a broad zone. Amacrine cell terminals branch on d10, and contact ganglion cell dendrites, forming a further band of synaptic junctions (Coulombre 1955). Coated vesicles, interpreted to be precursors of synaptic vesicles, are found in the inner plexiform layer by d12 (Armengol et al. 1979). Conventional synapses and synaptic vesicles are described for d13; 1 day later synapses with ribbons are found. The number of synaptic vesicles and ribbons increases then, and by d19, the inner plexiform layer is well developed (Sheffield and Fishman 1970; Hughes and LaVelle 1974, 1975).

The receptor cell terminals are reported to be invaginated by dendrites as soon as d10, while only ribosomes are formed within them (Shiragami 1968). Two days later, the first synapses between receptor terminals and bipolar cells (Meller and Tetzlaff 1976) and, on d13, triads and E-PTA positive sites are formed (McLaughlin, 1976a), whereas dendritic projections from bipolar cells are reported to extend toward the outer plexiform layer no sooner than d14 (Yew and Meyer 1975). On this day, punctate junctions, but no synaptic specializations, are present in the outer plexiform layer (Hughes and LaVelle 1974, 1975). On d15, an enhanced formation of vesicles and synaptic lamellae is observed (Shiragami 1968). By the technique of freeze fracture, irregularly dispersed particles, which are assumed to be synaptic vesicles, can be shown on the outer surface of the receptor terminals and of dendritic invaginations (Meller and Tetzlaff 1977); by the E-PTA technique specific substances can now be revealed at postsynaptic sites and in the synaptic cleft (McLaughlin, 1976a). On the following day, the particles found on dendritic invaginations after freeze-fracture preparation become more numerous and form aggregates (Meller and Tetzlaff 1977). Synaptic sites of the receptor cell terminals are judged to be mature by d18. The aspect of the terminals is adultlike (Meller 1968).

As early as d4, future receptor cells are found to form apical protrusions containing a centriole and a cilium (Hanawa et al. 1976). On d8, when a single ciliary protrusion is observed (Mishima and Fujita 1978), cone inner segments (Ueno 1961) are identified, which later on produce oil droplets. On the following day, the inner segments are spherical in shape and produce microvilli (as the adjacent Müller cells do); within their cytoplasm, an apical accumulation of mitochondria is observed as well as ER, a Golgi zone, centrioles near to the apical cell membrane, and polyribosomes, which are aggregated to form rosettes; at the level of the junctional complexes, microtubuli are oriented parallel to the longitudinal axis; desmosomes connect them to neighboring cells. On d12, microvilli are mainly found at the lateral part of the inner segments. The arrangement of the organelles is similar to that of the mature receptor cells; an ellipsoid is formed and in the myoid, ribosomes as well as rough and smooth ER are present. On d14, many vesicles of the sER near to the nucleus are regarded as precursors of the paraboloid. On the surface of the inner segments, besides elongated microvilli an isolated cilium is formed on d15. A network of vesicles and tubuli, which are filled with electron-dense material, is a further step in paraboloid formation. On d16, the inner segments elongate and become ovoid in shape; there are less microvilli now; the centrioles lie close to the pigmented epithelium, and protruding cilia are surrounded by microvilli, which will become calycal processes (Meller and Breipohl 1965; Olson 1973, 1979). Polyglucose and particulate glycogen as well as PAS-reactive glycogen (Yew 1976) and a paraboloid (Yoneyama 1932, cited after Yew, 1976; Olson 1972) are present in inner segments on d17 (Amemiya and Ueno 1977). It has a spheroid shape and is formed of sER vesicles, which are connected with rER; glycogen is found in it only from now on (Meller and Breipohl 1965). The paraboloid is distinct in cone cells by d18 (Rebollo 1955); in rods a body called "hyperboloid" appears at the same time (Yew 1976).

The outer segment is represented by a ciliary process on d12 (Leplat 1914; Meller 1968; Hanawa et al. 1976), and on d13, the first short disks are formed without being continuous (Govardovsky and Kharkeevich 1965; Meller 1964), but the developing outer segments are also characterized by vesicular and tubular structures (Hanawa et al. 1976). On d14, the cone outer segments have not surpassed the stage of a ciliary outgrowth and some membrane infoldings (Ueno 1961; Hanawa et al. 1976). Disks of

13

rod and cone outer segments are regularly produced from d15 on by infoldings of the plasma membrane (Govardovsky and Kharkeevich 1965).

Small outer segments are discernible in scanning electron microscopy on d16 (Meller and Tetzlaff 1976). On this, as well as on the following day, the size of the outer segments and the number of disks increase (Hanawa et al. 1976). At first the disks are regularly oriented; after apical movement, this orientation is lost. The outer segment still increases rapidly in length on d18. Definitive structure is attained when disks lose the contact to the plasma membrane (Govardovsky and Kharkeevich 1965, 1966). In the scanning electron microscope, inner and outer segments have their mature shape by d20 (Olson 1979).

2.3.5 Neurochemical Differentiation

As a first representation of specific substances, acetylcholinesterase is found in the centralmost ganglion cells on d4. Two days later, all ganglion cells and the first recognizable amacrine cells contain AChE, localized in the cytoplasm or on the cellular surface. Also in the inner plexiform layer, appearing at that time, a diffuse but high AChE activity is found (Shen et al. 1956). On this day, cholineacetyltransferase has also been found to be present, but at a low level (Crisanti-Combes et al. 1978). The observation that at this age binding sites for α-bungarotoxin are present in the retina (Vogel and Nirenberg 1976) and a ligand for muscarinic acetylcholine receptors is found (Sugiyama et al. 1977) fits well to these reports. An increase in AChE begins on d7 and lasts up to d18 or 21, the day of hatching (Lindeman 1947; Ramirez 1977). Up to d11, it is paralleled by an increase in synthesis and storage of acetylcholine and an increase in choline acetyltransferase activity, stated to be 1000-fold (Bader et al. 1978) or to run from 20–300 pmol/min/mg protein (Crisanti-Combes et al. 1978). On d8, histochemically revealed AChE in the inner plexiform layer is found to correspond to "synaptic bands" topographically (Shen et al. 1956). Accordingly, acetylcholine receptors on d9 are found to be localized mainly in the inner plexiform layer, as visualized by the binding of ^{125}J-α-bungarotoxin (Vogel and Nirenberg 1976). From d11 up to d21 (hatch), a phase of irregular synthesis and storage of ACh is noted, paralleled by a distinct increase in α-bungarotoxin binding sites, and a sixfold increase in velocity of the high-affinity uptake of the transmitter. Contrary to what has been found up to d11, no or only a slight enhancement of cholineacetyltransferase activity of 300–350 pmol is found; therefore, an increase in the number of available sites must be assumed (Lindeman 1947; Wang and Schmidt 1976; Bader et al. 1978; Crisanti-Combes et al. 1978). On d13, the concentration of α-bungarotoxin receptors has increased tenfold since d9, but the main bulk, namely 80%, will be synthesized during the following days, mainly in the inner plexiform layer (Vogel and Nirenberg 1976). In this layer, two distinct bands of muscarinic ACh receptors can be localized on d13, and a ligand for these receptors is present in the concentration of 320 fmol/mg retina tissue (Sugiyama et al. 1977). A further increase in cholineacetyltransferase activity of 300–350 pmol/min/mg (Crisanti-Combes et al. 1978) is accompanied by the adult aspects of four bands of acetylcholinesterase and of synapses in the inner plexiform layer as well as by the appearance of AChE activity in receptor cell inner segments, where the myoid will be formed (Shen et al. 1956). On d17, α-bungarotoxin is bound to the inner plexiform layer in distinct bands, and for the first time it is also bound in the outer plexiform layer (Vogel and Nirenberg 1976). For d19, a sudden increase in acetylcholine content of 7.8 to 12.5 γ/g retina is noted. No further change in cholinesterase is found after d20 (Lindeman 1947). At that time this enzyme has attained its final distribution in the ganglion, bipolar and receptor cells, as well as in the plexiform layers (Macri and Rebollo 1965). At the time of hatching, the increase in acetylcholine and acetylcholinesterase, as well as the increase in the numbers of binding sites for α-bungarotoxin, have come to an end; the latter have attained maximum concentration and are reduced to 1.5 fmol up to d3pp, while the transmitter 5 h after hatching shows a sudden decrease of 12.5 γ to 5γ/g retina (Lindeman 1947; Vogel and Nirenberg 1976; Wang and Schmidt 1976; Ramirez 1977). The adult number of binding sites for α-bungarotoxin is attained on d7pp (Wang and Schmidt 1976).

Biogenic amines seem to be represented before d12, when the monoaminoxidase is found to be faint and just begins to increase. On d18, this enzyme has attained its peak and a period of maximum activity follows, which lasts up to d7pp. The adult level is attained by a decrease from d7 to d14pp (Suzuki et al. 1977). On d6, ornithindecarboxylase, a characteristic enzyme of GABA production, shows high activity, but decreases sharply until d12, whereas glutamic acid decarboxylase, which stands for another GABA pathway, is low (de Mello et al. 1976). Perhaps in connection with this fact, GABA is accumulated in the retina by two systems of different kinetics on d9 (Tunnicliff et al. 1975). Its concentration has reached half of its maximum on d11 and further increases up to the stage of hatching, but at the beginning of d18, nearly all GABA is produced from glutamic acid (de Mello et al. 1976). Between d17 and d20, GABA differs from most other free amino acids in that its concentration increases (Pasantes-Morales et al. 1973). From the two systems for the uptake of GABA, only the one with low affinity is present at 3Wpp (Tunnicliff et al. 1975).

Glutamine is one of the amino acids which along with GABA increase after d17 (Pasantes-Morales et al. 1973); the level of glutamotransferase activity corresponds to the adult one on d1pp (Rudnick 1959). Cysteine sulphinate decarboxylase, indicating taurine synthesis, increases threefold between d10 and d7pp; its main increase can be found between d11 and d17 (Pasantes-Morales et al. 1976). On d15, hydroxyindole-O-methyltransferase is demonstrated for the first time, an enzyme which is characteristic of indoleamine metabolism (melatonine synthesis) in the pineal organ. The increase is still continued on d21, but at a lowered rate, and even at 2Wpp, its activity has attained only half of its adult level after a 30-fold increase since d15 (Wainwright 1979).

Cyclic AMP is detected in the retina on d6 in a concentration of 10 pmol/mg protein. An increase, which is normally found between d16 and 18, can be elicited on d7, when dopamine is administered. At the hatch, an abrupt rise to 20 pmol is noted. The concentration of cGMP is 30-fold lower, but shows an increase of up to 4.5 pmol at hatch (de Mello 1978; Fletcher and Chader 1978). The cGMP-phosphodiesterase decreases specifically on d8, but rises from d14 up to hatch. This rise is paralleled by a decrease in the cGMP level (Chader et al. 1974; de Mello 1978).

While it may be supposed that early GMP levels are involved in events leading to cell division, the increase of cGMP and cAMP at the onset of visual function suggests a role in neural processes.

Adenosine triphosphatase (apyrase) reaches a first peak on d8 and a second peak on d15 (Coulombre 1955). In the outer segments, ATPase activity increases at the time of hatching (Yew et al. 1975).

On d8, the concentration of phospholipids, N-acetylneuramine acid, and of the ganglioside G_{D1a} increases, while G_{D1b} and G_{D3} drop to adult level until d11, when the first phase of neuramine acid and ganglioside synthesis is finished (Dreyfus et al. 1975). Nevertheless, the activity of enzymes indicating ganglioside (lactosylceramide) synthesis is highest on this day (Dreyfus et al. 1977). The second phase of increase, about d16, is characterized by neuraminic acid, phospholipids and the gangliosides G_{M3} and G_{M1} while G_Q and G_{Q1} decrease. A third phase does not begin before the 3rd Wpp (Dreyfus et al. 1975).

Retinol receptors are present in the retina and in the pigmented epithelium on d11, before outer segments are formed (Wiggert and Chader 1975), and for d14, meta-rhodopsin is reported to be localized in the retina whereas no contact between the neural retina and the pigmented epithelial layer seems to occur; the rhodopsin content now starts with a hyperbolical increase. Metarhodopsin II is found 1 day later (Mason and Bighouse 1975). On d19, rhodopsin is detected by spectrophotometry and may be extracted. Maximal value of its amount is attained on d4pp (Witkovsky 1963).

In the receptor cell inner segment of d5, a golden-yellow pigment which is similar to that of the oil droplet in the double cone main element can be shown spectroscopically. There is an increase in oil droplet material on d11, but this is not yet visible (Cooper and Meyer 1968). For the first time colorless oil droplets can be seen light microscopically on d13 (Coulombre 1955). Two days later, the whole retina content of red astaxanthin as indicative of single cone inner segments increases rapidly (Cooper and Meyer 1968), and is paralleled by an increase in the size of astaxanthin droplets as well as by a concomitant reduction in the number of these droplets. On d17, astaxanthin is found in inner segments all over the retina (Coulombre 1955).

2.3.6 Differentiation of Specific Activity

As early as d7 (Hanawa et al. 1976) or d8 (Geoto 1950) a small positive potential is found to be evoked by light stimulation. Since this happens long before the differentiation of specific retinal structures, it is assumed that the pigmented epithelium is responsible for this reaction. By d16 the PII wave can be discerned as the first component of the electroretinogram (Goto 1950). One day later, not only a typical early receptor potential is found, which subsequently gains in size (Hanawa et al. 1976), but the first true ERG consisting of a_1-, a_2-, b-, and d-waves (Rager 1979). This author stresses the variation between individuals: similar reactions were found in other specimens not earlier than d18. In addition a c-wave is found on d18 (Ookawa 1971a, b), when Hasama (1941) recorded an embryonic electroretinogram. By this day, on and off responses as well as individual responses to single light flashes also appear (Peters et al. 1958). On d19, when still the same components of the ERG are reported, the latency is twice as large as that of the adults, and the size is smaller (Goto 1950; Blozovski 1971a; Witkovsky 1963).

At that stage there are the first reactions to flicker light, similar to those of the hatchling (Peters et al. 1958). It should be mentioned here that the first behavioral reaction to light stimulation was found between d18 and d20 (Kuo 1932).

On d20, a postnatal ERG is being formed (Garcia-Austt and Patetta-Queirolo 1961), and the b_3-component appears (Witkovsky 1963). There is a behavioral reaction to 1 min of light (Oppenheim 1968). Up to d21, the ERG has constantly increased in size, the b-wave showing a sudden threefold increase, while the c-wave does not change its size of 100 μV (Goto 1950; Witkovsky 1963; Ookawa 1971b; Ookawa and Takahashi 1971). On and off responses to sustained stimuli and individual responses to single light flashes are recorded in the eye and in the tectum (Peters et al. 1958). A maximum in ERG response on d5pp is followed by a decrease (Goto 1950), while a growth phase of the c-wave begins, lasting from d6pp to d10pp (Ookawa and Takahashi 1971). On d7pp, the initial latency of the ERG equals that of the adult retina (Blozovski 1971a).

A migration of the epithelial cell pigment can be observed as a result of an illumination by sunlight on d4pp. It attains its maximum at 2Mpp (Hasama 1941).

2.4 Differentiation of the Retina in the Mouse, *Mus musculus*

2.4.1 Proliferation and Migration

Mitoses are regularly found on d11 and earlier; at this time, ganglion and amacrine cells begin with their last cell cycle, which will be completed on d17 (Sidman 1961a) and the production of cone receptor cells seems to have started (Carter-Dawson and LaVail 1979). The first ganglion cells finish their migration on d12 (Hinds and Hinds 1974). On d13, many horizontal cells leave the mitotic cycle (Sidman 1961a) and the first rod cells are generated. Cone cell production reaches a peak on d14, followed by a sharp decline. From d17 onward, there is no further production of cone cells. Rod cell generation has its peak on the day of birth and comes to an end on d5pp (Carter-Dawson and LaVail 1979). Müller cells and bipolar cells complete their last cell cycle on d4pp, at least those of the most central region (Blanks and Bok 1977). There are no further mitoses after d6pp (Sidman 1961a; Caley et al. 1972).

2.4.2 Tissue and Cell Shape; Layering

On d11, the retina is still in its optic cup stage (Sidman 1961a) and a nerve-fiber layer will be identified from d12 on (Hinds and Hinds 1974). With d14, a decrease in inner neuroblast layer thickness begins, which will last up to d41pp and is probably due to cell degeneration (Blanks and Bok 1977). One day later, only ganglion cells and their neurites are layered. In the region of the inner neuroblasts, precursor cells of ganglion, amacrine, and horizontal cells are found. Among horizontal and receptor cells, others have been discovered which lack long vitreal processes and will presumably be interplexiform cells (Hinds and Hinds 1976, 1978). On d16, the inner plexiform layer appears (Bhattacharjee and Sanyal 1975), which on the following day is seen as a distinct region, made up of mainly tangentially orientated processes (Hinds and Hinds 1978).

At the time of the birth, the retina consists of 16–20 rows of cells, the ganglion cell layer comprising two rows (Sorsby et al. 1959). Now the amacrine cells, too, are layered (Olney 1968a) and receptor cell inner segments surpass the outer limiting membrane (Caley et al. 1972). An outer plexiform layer is recognized in electron microscopic preparations by d4pp (Olney 1968a), under the light microscope on d6pp (Orr et al. 1976), and 1 day later it is found to consist of neurites and dendrites. At this time some receptor cells are considered to be degenerating (Caley et al. 1972). From d5pp onward, the inner nuclear layer decreases by 50% up to d15pp, while the inner plexiform layer increases (Fisher 1979). In C3H mice, degeneration in the retina is found on d11pp; on d15pp a remarkable loss of nuclei in the receptor cell layer is observed, which has been reduced to a single row by d17pp (Zimmermann and East-ham 1959).

2.4.3 General Differentiation

Along the vitreal as well as along the scleral border of the optic cup, the enzyme dia-phorase, taken to indicate enhanced differentiation, can be demonstrated for the first time on d12. On the following day, its activity rises in the whole retina, especially in the ganglion cells. On d16 it is found within these, mainly vitreal to the nuclei, on the day of birth also scleral to them, and increases at least up to d4pp (Bhattacharjee 1977).

The region where the plexiform layer will arise is characterized by the occurrence of carboxylesterase (Bhattacharjee and Sanyal 1975) on d13. Desmosomes can be ob-served in this layer on d1pp. In the horizontal cells diaphorase is of medium intensity on the day of birth and increases thereafter (Bhattacharjee 1977). In the future re-ceptor cells, the radiosensivity (LD_{50} up to d13: 3800 rad) will be reduced from this day on (Lucas 1961). Diaphorase activity is localized at the apical border of receptor cells on d14pp, in the inner segments not earlier than d7pp, when it is also found in bi-polar cells (Bhattacharjee 1977).

A faint PAS and alcian blue reaction can be obtained in receptor cells on d9pp, in-dicating the production of polysaccharides. Five days later, outer segments become in-creasingly PAS-positive and are surrounded by an alcian blue-positive coat. This in-crease lasts at least until d30pp (Zimmermann and Eastham 1959). The cells of the pigmented epithelium display PAS-positive material on d5pp (Karli 1961). By d30pp the amount of the latter is reduced (Zimmermann and Eastham 1959).

On d5pp, retina cells are no longer sensitive to cyclophosphamide, as it was the case before (Foerster and Lierse 1975).

2.4.4 Specific Structural Differentiation

On d12, the first ganglion cells start a visible differentiation by the production of axo-nal sprouts, becoming unipolar (Hinds and Hinds 1974). On the following day, den-drite-like processes are formed (Hinds and Hinds 1978). The first amacrine cell proc-esses are found on the day of birth, when they already form immature synapses. True synaptic junctions in the inner plexiform layer are reported for d3pp, when ganglion and amacrine cells also form subsurface cisterns (Olney 1968a, b). The synaptic junc-tions are of the conventional type and show the first rate of increase ($0.4/1000\ \mu m^3/h$) up to d10pp (Fisher 1979). On d10pp conventional synapses have attained a density of $85/1000\ \mu m^3$, which equals 49 synapses per amacrine or interplexiform cell, and mature at an enhanced rate in the inner plexiform layer; now the first serial synapses and bipolar cell terminals as characterized by synaptic lamellae appear too (Olney 1968a, b; Fisher 1979). Between d1pp and d15pp, the rate of conventional synapse increase rises to $1.2/1000\ \mu m^3/h$, that of the bipolar cell synapses to $0.4/1000\ \mu m^3/h$, with their small ovoid ribbons becoming longer (Olney 1968b; Fisher 1979). On d15pp, the density of conventional synapses has reached the level of 223/1000 μm^3; this equals 252 synapses per amacrine or interplexiform cell, while only 45 ribbons per 1000 μm^3 or 49 per bipolar cell have been counted; they will, however, increase to 113/1000 μm^3 after d16pp (Fisher 1979). Conventional synapses of the inner plexiform layer are added at a reduced rate on d18pp. Then bipolar cell terminals are no longer discernible from adult ones and are no longer formed after d20pp (Olney 1968a).

The first synapses between horizontal and bipolar cells in the outer plexiform layer are reported for d10pp. As soon as d2pp, however, the first primitive receptor terminals are characterized by synaptic vesicles and synaptic lamellae and 2 days later they show a synaptic spindle (Olney 1968a, b). On d5pp, in addition to what has been found before, the receptor cell terminals acquire synaptic densities, asymmetric membrane thickenings, synaptic cleft material, and postsynaptic dense projections; that means that they possess everything to provide for mature synaptic sites. All invaginating dendrites stem from horizontal cells. Dyads as well as multiple synaptic lamellae are found on d6pp (Blanks et al. 1974a; Olney 1968a). On d7pp, the terminals show the typical adult triadic configuration (Olney 1968a) and even in rd mice with genetically programmed cell degeneration dyads are formed up to this date. But during the following days, no triadic configurations will appear in rd mice (Blanks et al. 1974a). (In view of the fact that dyads have been formed in an otherwise normal receptor terminal, one might ask whether it is really the receptor cell which causes the degeneration. It is the bipolar cell process which lacks in triad configuration. The subsequent degeneration of the receptor cell might be due to the lack of an induction normally exerted by the process.) By d10pp, the majority of receptor cell terminals show triads and synaptic spindles; they are fully developed by d12pp (Olney 1968a, b) or d14pp (Blanks et al. 1974a).

By d12pp the receptor cell inner segment is characterized by Golgi elements and ER as well as by an ellipsoid, which, however, is far from being adultlike (de Robertis 1956).

As a first indication of the outer segment, a ciliary stalk is noted to protrude from the inner segments on d1pp; and on d3pp, its apical part becomes balloon shaped. The first disks appear on d6pp (Olney 1968a). They only display a few intramembraneous particles, which are supposed to represent visual pigment (Olive and Recouvreur 1977). On the following day, the outer segments enclose 20 disks (Caley et al. 1972).

Parallel to an increase in the number of disks, an increase in the density of intramembraneous particles (= visual pigment?) can be found (Olive and Recouvreur 1977). On d11pp, the rate of outer segment disk production is higher than in the adult retina, while the rate of disk shedding is about 15% that of the adult; this results in a net length increase of the outer segments. A rise in the disk production rate to 1.6 times of the adult retina results in a period of a particular sharp rise in the outer segment length, which lasts up to d17pp (LaVail 1973). On d14pp, the length of the outer segment equals that of the inner segment (Sorsby et al. 1959), disks are regularly and densely packed, and the density of the intramembraneous particles has increased (Olive and Recouvreur 1977). The phase of enhanced outer segment growth comes to an end on d17pp, but the adult relation between outer segment disk production and disk shedding is not attained before the 4th Wpp (LaVail 1973).

2.4.5 Neurochemical Differentiation

As a first neurochemical agent, acetylcholinesterase is histochemically localized in ganglion cells at the date of birth. On d10pp, it is demonstrated in the inner plexiform layer (Bhattachorjee and Sanyal 1975).

On d1pp, an increase in GABA and glutamine as well as a sharp rise in glycine content is observed; the latter will last up to d8pp, when a rise of taurine begins. The glycine decrease comes to an end by d15pp, and a phase of glutamine decrease begins (Orr et al. 1976).

The activity of adenylcyclase amounts to 25 pmol/mg protein/min at d1pp and rises to 50 pmol on d10pp, which is about adult level. But it is only from d5pp on that it can be stimulated by dopamine. After an initial increase the content of cAMP shows no further change from d13 on (Lolley et al. 1974). The overall amount of ATPase is found to be 3 mM/g dry weight (cf., in the adult retina it is 8–10 mM/g). By d20pp C3H mice show a decrease in the amount of retinic ATPase (Lolley and Racz 1972).

2.4.6 Differentiation of Specific Activity

As the first indication of an electroretinogram, a small b-wave is recorded on d12pp; when high intensity stimulation is applied, a small a-wave is visible as well. Two days later, distinct a- and b-waves are recorded (Noell 1958). The first study on ERG development ever done revealed the first potential differences in the 14d-old neonatal mouse (Keeler et al. 1928). According to this study, the ERG has attained adult shape in 4 Wpp; the size, however, is still lower.

2.5 Differentiation of the Retina in the Rat, *Rattus norvegicus*

2.5.1 Proliferation and Migration

Mitoses can be found in the scleral zone of the retina at the central and peripheral regions between d13 and d7pp (Detwiler 1932; Zavarzin and Stroeva 1964; Braekevelt and Hollenberg 1970; Kuwabara and Weidman 1974). The duration of the cell cycle in the neonatal rat is 28 h (Denham 1967). On d3pp, a decline in mitotic activity begins, and on d7pp there are no mitoses at all (Detwiler 1932; Weidman and Kuwabara 1969; Braekevelt and Hollenberg 1970) or only at the ora serrata (Berkow and Patz 1964).

2.5.2 Tissue and Cell Shape; Layering

On d11, the formation of the optic cup is in progress (Braekevelt and Hollenberg 1970), but at the same time the onset of morphogenetic cell death has been reported (Silver 1976). One day later, the formation of the optic cup is completed (Kuwabara and Weidman 1974), the fissura optica is formed, and the rate of necrotic cells decreases (Silver 1976). The whole retina then is an undifferentiated epithelium with an indication of a nerve-fiber layer (Berkow and Patz 1964). On d15, this layer is composed of numerous neurites (Kuwabara and Weidman 1974; Raedler and Sievers 1975). The innermost two or three rows of the neuroblastic layer have round nuclei, and by d16

20

or 17, layering starts in ganglion cells (Braekevelt and Hollenberg 1970; Kuwabara and Weidman 1974). The beginning of an inner plexiform layer is reported for d17 (Raedler and Sievers 1975) or d18/19, by inward migration of ganglion cells (Braekevelt and Hollenberg 1970). At this time it is composed of only short and thick cytoplasmic protrusions, but shortly before birth, more and more different processes appear (Raedler and Sievers 1975). On d1pp, the ganglion cell layer is still made up of several rows of cells (Schmidt 1973), and on d2pp, they are the only cells to be layered, but amacrine cells show signs of beginning differentiation (Denham 1967).

A receptor cell inner segment is visible in Golgi preparations on d1pp (Morest 1970). At d4pp, when bipolar and amacrine cells may be distinguished from each other, an outer plexiform layer appears (Detwiler 1932; Schmidt 1973; Raedler and Sievers 1975). It is distinct on d5pp (Braekevelt and Hollenberg 1970) and at light microscopic level on d6pp (Detwiler 1932). Up to d9pp, both plexiform layers have increased in thickness and are now represented even in peripheral regions (Braekevelt and Hollenberg 1970).

The Müller cells, which on d15 begin to extend horizontal processes and on d3pp separate receptor cells, fill all intercellular spaces in the inner plexiform layer on d10pp, which therefore appears more compact (Weidman and Kuwabara 1968; Kuwabara and Weidman 1974). Though bipolar cells are said to lag behind (Raedler and Sievers 1975), the light microscopic aspect of the retina is now adultlike (Berkow and Patz 1964). On d21pp, the retina displays regional differences in the pattern of ^{3}H-uridine labeling between the anterior and the posterior pole (Yew 1979).

2.5.3 General Differentiation

On d14, all retinal cells contain a lot of polysomes and single free ribosomes. In the ganglion cells, the volume of which increases, cisterns of the endoplasmic reticulum develop, as well as lots of mitochondria and conical protrusions directed to the vitreous. Müller cells show sER, a Golgi zone, and lipid inclusions; they are supposed to be phagocytotically active (Kuwabara and Weidman 1974). As early as d13, ventricular cells are characterized by the accumulation of mitochondria, by microtubuli, and by centrosomes. A number of growth cones is found to sprout from the bipolar-shaped cells (Raedler and Sievers 1975). On d16, all neuroblasts contain polysomes and a Golgi apparatus (Spira 1974). By d20, a polarization of the ganglion cells is expressed by the fact that mitochondria, Golgi apparatus dense bodies, rER, and neurofilaments appear concentrated in the dendritic part of the cells.

From the cells of the inner nuclear layer axon hillocks sprout. They contain mitochondria, Golgi lamellae, and rER; mitochondria and ER are also found in the now recognizable horizontal cells. In future inner segments, ribosomes, mitochondria, and centrosomes are present on d18 and still on the day of birth; microtubuli are located vitreal to the nucleus (Weidman and Kuwabara 1969; Raedler and Sievers 1975; Galbavy and Olson 1978). One day later, there is a microvillous contact to the pigment cells (Morest 1970).

Cellular tight junctions connect developing receptor cells on d13 as well as at birth (Weidman and Kuwabara 1969; Kuwabara and Weidman 1974; Raedler and Sievers 1975). On d5pp, the junctions in the inner plexiform layer may be nonsynaptic as well as synaptic precursors (Weidman and Kuwabara 1969).

While alkaline phosphatase is not found at all, acid phosphatase can be localized in ganglion cells on d1pp. Three days later, this enzyme is also demonstrated in amacrine cells, and in bipolar and cone receptor cells on d7pp. While it also appears in rods on d11pp, it is still in its phase of increase in bipolar and amacrine cells (Schmidt 1973).

On d13, the inner and the outer nuclear layer are bordered by histochemically detected enzymes, which, in the following, are taken as indicative of a general metabolic activity and enzymatic differentiation (Hellström 1956, cited after Coulombre, 1961; Berkow and Patz 1964; Utermann 1964; Wolff 1969). On d17, three layers of oxidative enzymes (succinic dehydrogenase, lactic dehydrogenase, TPN and DPN diaphorase) are found in the retina and the pigmented epithelium. In ganglion cells, NADH-diaphorase is concentrated in a cap at the nerve-fiber site; in the nerve-fiber layer cytochemical reactions are weak. The receptor cells show a reaction which is similar to that of the ganglion cells.

On the day of birth, there is a strongly LDH- and diaphorase-positive cap in the ganglion cells, which is orientated toward the inner plexiform layer; now bipolar cells, too, are LDH-positive in a cap which is directed toward the vitreal part of the retina. In the inner plexiform layer, a moderate activity of succinic dehydrogenase and glucose-6-phosphate dehydrogenase is found; the former is now stronger in the receptor cells. One day later, SDH appears in inner segments and is still localized there on d4pp when horizontal cells show LDH. On d7pp, the oxidative enzymes show high activity in the rods and the pigmented epithelium, but less activity in the plexiform layers, and weak activity in the nuclear layers.

On d9pp, the outer plexiform layer is characterized by a high activity of diaphorase, LDH, SDH, and glucose-6-phosphate-dehydrogenase. Additionally, SDH is found in the inner plexiform layer and in the ellipsoid. On d10pp, the oxidative enzymes show high activity in the pigmented epithelium, the outer and inner segments, the ganglion cells, and all neurite layers. Synchronously, the activity of malate dehydrogenase and glycerinaldehyd-3-phosphate dehydrogenase begins to rise. Two days later, G-6-PDH has become distinct in the inner plexiform layer, showing a stratification similar to unspecific esterases. On d14pp, LDH, MDH, and DPN are highly active in the pigmented epithelium, the ellipsoid, the plexiform layers, and the ganglion cells. SDH is moderately active in the outer segments as the TPN is in the pigmented epithelium, the outer plexiform layer, and in ganglion cells and their neurites. On d15, the LDH is similar to that of the adult retina. The activity of 5-nucleotidase can be demonstrated histochemically after d18; on d4pp, it is present in receptor cells and on d8pp in the zone of the outer plexiform layer. Unspecific esterases are found in the ganglion cells not earlier than d9pp; and by d21pp their amount and distribution in these cells is adultlike.

From d3pp to d10pp, the rate of anaerobic glycolysis remains constant. It is followed by an increase between d10pp and d14pp; there is no further change. From d5pp to d8pp, a constant rate of aerobic glycolysis is found. It is followed by an increase between d12pp and d30pp; then there is no further change (Graymore 1959). It becomes clear that metabolic activity is lower in young than in adult retinae (see also Cohen and Noell 1960). The respiratory capacity of the retina is doubled within the 2 weeks following d7pp, but an increase will be found until d30pp (Graymore 1960; Utermann 1964).

Free amino acids by d30pp have not yet attained adult level and are still subject to fluctuation (Macaione et al. 1974).

On d1pp, ganglion cells and the inner plexiform layer are seriously damaged by irradiation with 500 R, but receptor cells become pycnotic then and degenerate (Schmidt 1973). While for d3pp, the retina as a whole is said to have become more sensitive to 200 R than it was before (Mock et al. 1975), for d4pp no effect of 500 R on ganglion, amacrine, and receptor cells is reported; only the bipolar cells become pycnotic (Schmidt 1973). On d7pp, irradiation with 500 R is entirely without damaging effect (Schmidt 1973), but receptor cells are still sensitive to a laser beam (Yew and Chang 1977).

After d1pp, the effect of cyclophosphamide (the production of rosettes of degenerating receptor cells) will be reduced, and from d5pp on, it is completely abolished (Foerster and Lierse 1975).

2.5.4 Specific Structural Differentiation

The first ganglion cells are found on d15 (Raedler and Sievers 1975). By d17, the ganglion cells develop conical protrusions with ER, which will develop into dendrites (Kuwabara and Weidman 1974). Amacrine cell processes follow after d18; and on d3pp, ganglion cell dendrites ramify in the inner plexiform layer (Raedler and Sievers 1975), where perhaps synaptic precursors are found 1 day later. A true differentiation of this layer, however, is not reported before d8pp, when synaptic vesicles and synaptic ribbons appear (Weidman and Kuwabara 1969). By d10pp or d11pp, there are synaptic junctions, the number of which increases (Spira 1974; Raedler and Sievers 1975).

On d4pp, increasing numbers of Müller cells are found to contain glutamine synthetase, and on d7pp, their processes are detected in all layers of the retina (Riepe and Norenberg 1978).

Processes grow out of the horizontal cells and a receptor cell terminal develops, characterized by synaptic sites, vesicles, and ribbons on d6pp (Weidman and Kuwabara 1969). But as soon as the day of birth, mitochondria migrate into the receptor cell inner segment. By d6pp, mitochondria, polysomes, ER, and Golgi lamellae can be detected (Bok 1968). Two days later, the inner segmet has elongated, and an ellipsoid is well developed (Galbavy and Olson 1978).

On d5pp or d6pp, the first disks are formed by evagination of flat saccules. The number of those increases then (Bok 1968; Weidman and Kuwabara 1969). In the scanning electron microscope, the outer segment can be seen as a bulbous expansion by d8pp (Galbavy and Olson 1978). On d11pp a phase of enhanced growth begins, and on d12pp the first phagosomes are produced (one to three per 200 μm of outer segment length) (Detwiler 1932; Tamai and Chader 1979). On d14pp, after having completed the phase of rapid growth the rods have nearly reached adult size (Berkow and Patz 1964) and are adultlike as far as ultrastructural aspects are concerned (Raedler and Sievers 1975; Bonting et al. 1961). One day later, they show a length of 11–13 μm (Dowling and Sidman 1962); a circular rhythm of phagosome production begins (Tamai and Chader 1979). The adult length is finally reached on d21pp (Berkow and Patz 1964; Weidman and Kuwabara 1969), but phagosome production attains its final level of 30–40/200 μm not earlier than d30pp (Tamai and Chader 1979).

2.5.5 Neurochemical Differentiation

Acetylcholinesterase is found in the inner plexiform layer on d7pp (Spira 1974) or d8pp, and will be localized in bands (Wolff 1969). On d21pp, it is still found in this region, but not in the outer plexiform layer (Spira 1974).

Even at birth, there is a rising level of GABA concentration, which is followed by a sharp increase on d15pp; the final state is reached on d30pp (Macaione et al. 1970).

Beginning on the date of birth, glutamate increases threefold up to d30pp and glutamine synthetase 20-fold (Macaione and Cacioppo 1971). The latter enzyme is said to occur only in Müller cells (Riepe and Norenberg 1978). It increases between d9pp and d17pp, when it attains a 25-fold level; after this day there is no further change (Chader 1971; LaVail and Reif-Lehrer 1971). While there is a graded decrease in the concentration of glutamine after d7pp, a rise in glutamate dehydrogenase begins, which will result in an 11-fold increase up to d60pp (Macaione and Cacioppo 1971). On d15, two enzymes involved in the glutamate metabolism have reached maximum activity (Macaione et al. 1973).

Between d6pp and d10pp, the uptake of ^{14}C-glycine from the surrounding medium increases by 50%; then it remains constant, while the incorporation into retina protein decreases at this time and after d12pp (Reading 1965).

The first neurochemical substance to appear is hydroxy-indole-0-methyltransferase, an enzyme of the melatonin metabolism, which had previously been found in the pineal organ only; it can be demonstrated on d17. By d30pp, its activity has reached a plateau (15.9 and 12.1 pmol/melatonin/h in the male and the female respectively) (Cardinali and Rosner 1971).

Cysteine sulphinate decarboxylase, which is active in the synthesis of taurine, increases between d13pp and d30pp, especially up to d20pp (Pasantes-Morales et al. 1976).

At birth, the activity of adenylate cyclase begins to increase, and after d25pp it remains at adult levels; it is mainly localized in the inner nuclear layer (Farber and Lolley 1977). On d7pp, the amount of phosphate in the retina is 6.2 µg/h/mg wet weight. Of this, 3.2 µg are attributable to Na^+-, K^+-activated ATPase, which is less than half the adult value (Towlson 1964).

Rhodopsin is first detected on d6pp, and is soon characterized by an increase. On d12pp, it can be extracted in significant amounts. Two or three days later, the amount of visual pigment is half the maximum. By d21pp, the rhodopsin content in the whole retina reaches a plateau, but the adult level of 8×10^{-4} µmol per eye is attained only in the 6th Wpp (Bonting et al. 1961; Dowling and Sidman 1962).

2.5.6 Differentiation of Specific Activity (Fig. 2)

An a-wave with a size of only 50 µV appears as a first sign of an electroretinogram on d9pp. On d10pp, the threshold is 10 000 times that of the adult. The fast initial wave is then followed by a slow cornea-negative wave; the amplitude is 75 µV. Two days later, the rise of the ERG has increased, and in addition to the a-wave a positive b-wave appears, but is still rather insensitive. The ERG is completed by a prominent b-wave on d15pp, and its threshold is reduced to 1000-fold that of the adult. On d18pp, the ERG is almost adult in size (1000 µV) and the threshold is only tenfold adult. On d22pp, a- and b-wave level off and by d30pp, the a-wave reaches a plateau; the b-wave attains its

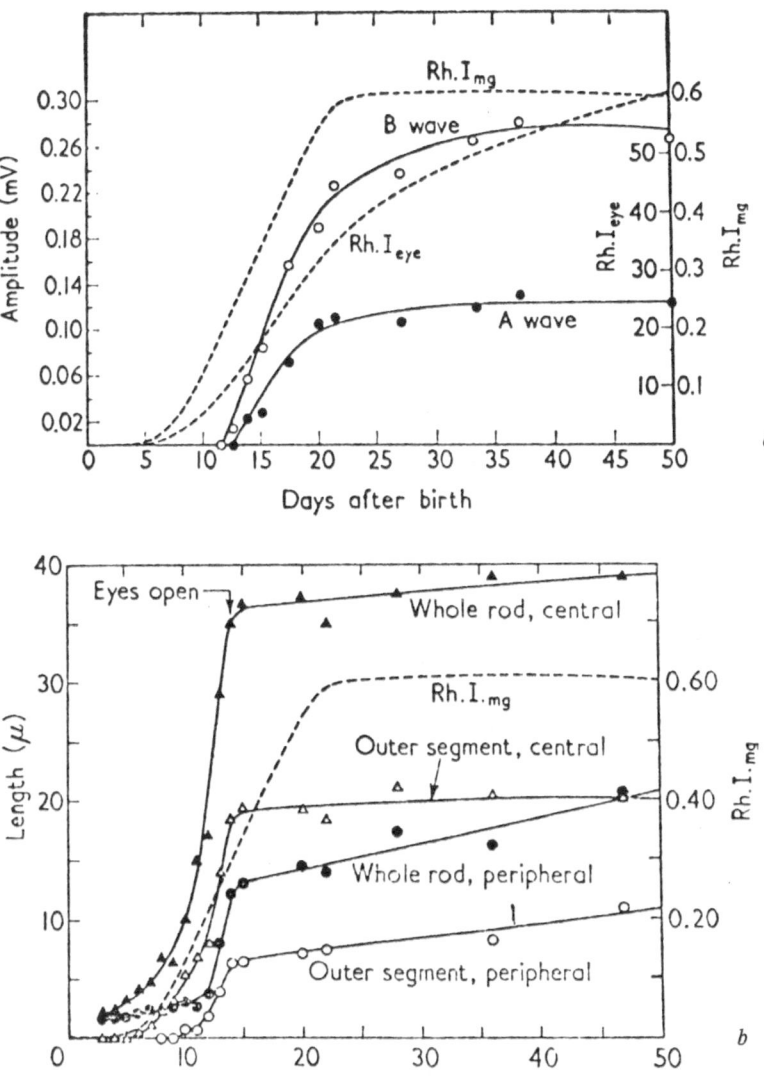

Fig. 2. *a* Correlation of electroretinogram amplitude development (a-wave, b-wave) in the postnatal rat with the amount of rhodopsin per mg of whole eye (Rh.I$_{mg}$) and rhodopsin content per eye (Rh.I$_{eye}$). *b* Correlation of the length of whole rods and outer segments in central and peripheral areas in the postnatal rat. (Bonting et al. 1961)

final level during 6Wpp, when the ERG increases from 1200 to 1500 μV (Bonting et al. 1961; Dowling and Sidman 1962; Hellner and Utermann 1965).

Before the first ERG can be demonstrated, an avoidance behavior is noted as the first reaction to light on d8pp (Detwiler 1932).

2.6 Differentiation of the Retina in Man, *Homo sapiens*

The development of the human embryo or fetus is differently staged, by days, weeks, months, or crown-rump-length. Discrepancies between authors are thereby increased. Another source of discrepancy is the difference between development processes in foveal and extrafoveal regions, the former showing partly an acceleration, partly a delay, in its development; not all authors explain from which part of the retina their data derive.

2.6.1 Proliferation and Migration

Mitotically active cells are found from d24 (2.5 mm) on (Seefelder 1910; Barber 1955; Duke-Elder and Cook 1963; Mann 1964; O'Rahilly 1975). At a length of 100—120 mm there are still some mitoses (Rhodes 1979), but after this stage, there will not be any more (Hollenberg and Spira 1973).

2.6.2 Tissue and Cell Shape; Layering

By d24 (2.5 mm) an optic vesicle has been formed; the future retina cells are a mere epithelium of elongated cells (Seefelder 1910; Duke-Elder and Cook 1963; O'Rahilly 1975). At 4W (4—4.5 mm) the retina consists of eight to nine rows of cells and is 0.09 mm thick (Mann 1964). At 8W (20—23 mm) adult thickness is attained and from now on the retina only grows by extension of its surface (Duke-Elder and Cook 1963). At 12W (60 mm), the macula and fovea, which, up to this time, were in advance of the rest of the retina, slow down and will lag behind the more peripheral regions (Duke-Elder and Cook 1963; Barber 1955). Layering is distinct at 26W (Horsten and Winkelmann 1962). At birth, the more peripheral regions of the macula and the nonmacular retina are well developed (Duke-Elder and Cook 1963).

A nerve-fiber layer is formed at 5W (5—7 mm) (Seefelder 1910). In the 9th W (30—31 mm) the optic chiasma formation is completed but in the central region of the macula a nerve-fiber layer is still lacking at birth (Duke-Elder and Cook 1963). At 300 mm, the ganglion cell layer comprises one row of cells only (Seefelder 1910). In the fovea, ganglion cells have nearly vanished by 4Mpp (Duke-Elder and Cook 1963).

Amacrine cells are distinguishable from ganglion cells by 10W (Mann 1964), and at 7M, there are numerous pycnotic nuclei among them (Kolmer 1936).

Before an inner plexiform layer is formed, the layer of Chievitz appears, which, in the 7th W separates an outer and an inner neuroblast layer (Mann 1964; Spira and Hollenberg 1973). Even at 8W, it also contains perikarya (Rhodes 1979, Fig. 3a). In the 9th W the layer of Chievitz disappears and a true inner plexiform layer (Fig. 3b) is being formed (Spira and Hollenberg 1973). The Chievitz layer persists in the macula and has its greatest extension in the 8th M, but at birth it is reported to have disappeared there (Duke-Elder and Cook 1963).

An outer plexiform layer appears at 10W (Spira and Hollenberg 1973) and is distinct under the light microscope in the 100—120 mm fetus (Barber 1955; Mann 1964).

Bipolar cells begin their layering extrafoveally at 12W (Seefelder 1910) and are well identified along with horizontal cells at 23W (Rhodes 1979). At 4 Mpp, they have nearly vanished in the fovea (Duke-Elder and Cook 1963).

Müller cells are the first ones to be recognized; they are found at 4W, along with the outer limiting membrane (Seefelder 1910). By 8W they are present in all existing layers (Rhodes 1979).

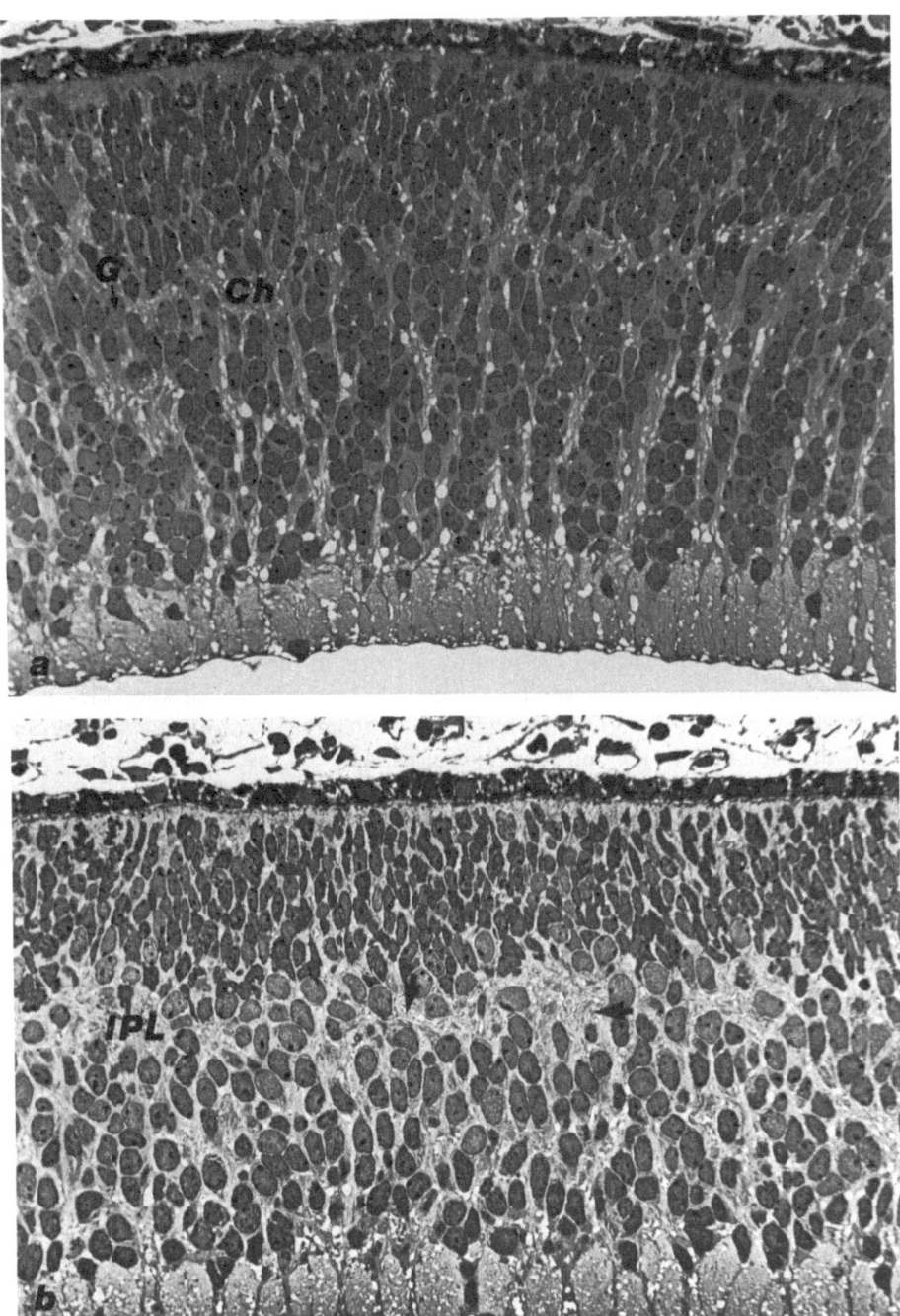

Fig. 3. *a* Formation of the layer of Chievitz (*Ch*) in the 41-mm human fetus. Ganglion cells (*G*) are still found in the outer neuroblastic layer. *b* 61-mm human fetus. By clearing of cellular nuclei the layer of Chievitz is transformed into an inner plexiform layer (*IPL*). The *arrow* indicates horizontal and vertical cellular processes. (Rhodes 1979) × 320

2.6.3 General Differentiation

At 5W, the cytoplasm of the ganglion cells is already argyrophilic (Rhodes 1979). In these cells, pyronine positivity, as indicative of Nissl substance, shows a marked increase at 4M (83 mm) (Nakayama 1957). In extrafoveal ganglion cells, the first Nissl bodies are seen in the 8th M (Seefelder 1910). In the cone inner segments, RNA is detected at 5M (150 mm) (Nakayama 1957).

At 6W the fusiform cells of the undifferentiated retina are interconnected by zonulae adhaerentes (Hollenberg and Spira 1973). In the 9th W, gap junctions, macula adhaerentia, and zonulae adhaerentes and occludentes are found in the retina and the pigmented epithelium (Fisher and Linberg 1975), but there are also junctions connecting retina and pigment layer (Hollenberg and Spira 1973). At 10W, terminal bars (Yamada and Ishikawa 1965) and desmosomes are reported for receptor cells; the latter are present up to a length of 100 mm; at 150 mm they have disappeared (Lerche 1963).

The first pigment granules appear in the pigmented epithelium as early as 5W (Seefelder 1910).

2.6.4 Specific Structural Differentiation

A scarce formation of ganglion cell neurites is found in the 7th W (Seefelder 1910); from now on they begin to invade the optic stalk. In the following week they reach the proximal part of the optic stalk, and some of them are near the neural tube; and at 8W, the formation of the optic chiasma begins. After its onset in the telencephalon, the myelination of the optic tract has advanced up to the optic chiasma at 7M, and at birth has reached the lamina cribrosa (Duke-Elder and Cook 1963).

The first ganglion cell dendrites are produced in the 9th W (Seefelder 1910). In the inner plexiform layer, synapses are reported to appear at 12W (Spira and Hollenberg 1973), while S.K. Fisher (1973, Development of the human retina, Abstract ARVO spring meeting, personal communication) did not find differentiated contacts even at 16W. According to both authors, conventional and ribbon synapses have developed at 5M. In the outer plexiform layer, synapses are found at 4M (Hollenberg and Spira 1973).

In cone terminals which have not extended yet, invaginations, synaptic junctions, synaptic vesicles, and synaptic lamellae are found; the latter appear either in connection with synaptic sites or float freely (Linberg and Fisher 1978). At 4M (83 mm), the cone cell terminals are characterized by synaptic vesicles and synaptic lamellae (Hollenberg and Spira 1973). At 100 mm, synaptic ribbons and synaptic vesicles are widely distributed. Synapses in rod terminals are found at 5M (Hollenberg and Spira 1973). Nearly at the same time, horizontal cell and bipolar cell invaginations form a triadic arrangement (Fisher 1973). In the 23W/200-mm fetus the cone terminals are spheroid and contain few mitochondria, few synaptic vesicles, and few dendritic invaginations; furthermore, ribosomes and tubuli are found. At 7M, the terminal has enlarged and comprises more dendritic invaginations and more synaptic vesicles (Yamada and Ishikawa 1965).

By 10W, receptor cell inner segments are recognized by an accumulation of mitochondria, polysomic ribosomes, ER, a Golgi system, and two centrioles. Cilia grow out of the latter (Yamada and Ishikawa 1965; Hollenberg and Spira 1973). Subsurface

28

cisterns are formed in adjacent cones at 12W (Linberg and Fisher 1978). Ciliary root-lets are seen at 100 mm (Hollenberg and Spira 1973). At 23W, mitochondria, ER, and ribosomes are localized in their definite positions. An ellipsoid is present at 7M, and in the 8th M inner segments are no longer different from adult cells (Yamada and Ishikawa 1965).

A primitive outer segment consisting mainly of ciliary tubules does not appear before 23W (Yamada and Ishikawa 1965), or 26W (Fisher 1973). At 7M, it is composed of irregularly orientated tubular structures (Yamada and Ishikawa 1965).

2.6.5 Differentiation of Specific Activity

It is in the 8th M that the first, incomplete electroretinogram can be recorded; it consists of an a- and a b-wave. Up to the 3rd Mpp, a continuous increase can be noted (Samson-Dollfus 1968). At birth, an electroretinogram is regularly found and has developed a- and b-components; the amplitude is lower than in the adult (Horsten and Winkelmann 1962; Samson-Dollfus 1968). Fourteen hours after birth, a positive off effect can be elicited (Fisher and Linberg 1975).

2.7 The Phases of Differentiation (Fig. 4)

In all the species considered here, processes of general and specific differentiation start when proliferation is still in progress. This is not unexpected, because the degree of maturity – as it has been known for a long time – is different in different zones of the retina. The most central part in the fundus of the eye (= posterior pole) is the most advanced, and the most peripheral part, near the ora serrata, lags behind most of all. New cells are added from the periphery (cf., e.g., Weysse and Burgess 1906; Müller 1952; Baburina 1956; Straznicky and Gaze 1971; Hollyfield 1972; Wagner 1972; Kunz and Wise 1974; Baburina et al. 1977; Carter-Dawson and LaVail 1979). Functional maturity, however, as demonstrated by the presence of ERG-like potentials, is normally attained only after the end of all mitotic processes in the retina. Only for *Xenopus* are mitoses reported for a very late stage, long after the structural and functional maturation have been achieved.

General differentiation begins as soon as the optic cup is formed and proliferation has come to an end in the centrally located cells. The onset of general differentiation is immediately followed by an onset of specific structural differentiation; in *Gallus* it is even preceded by this phase. Only in *Xenopus* does there seem to be a time lapse between general and specific differentiation, lasting, however, for some hours only.

Without exception, the first sign of specific differentiation is the outgrowth of ganglion cell neurites. In *Tilapia* and the mouse, other specific events follow considerably later. The further sequence of specific development will be treated in detail below.

The specific chemical differentiation begins in *Xenopus* earlier than the structural differentiation. In all other cases it starts later. Its most important part, however, is always paralleled by that of general and specific differentiation. In the chick, the cholinergic system is the first neurochemical system to appear. In *Xenopus*, the rat, and the

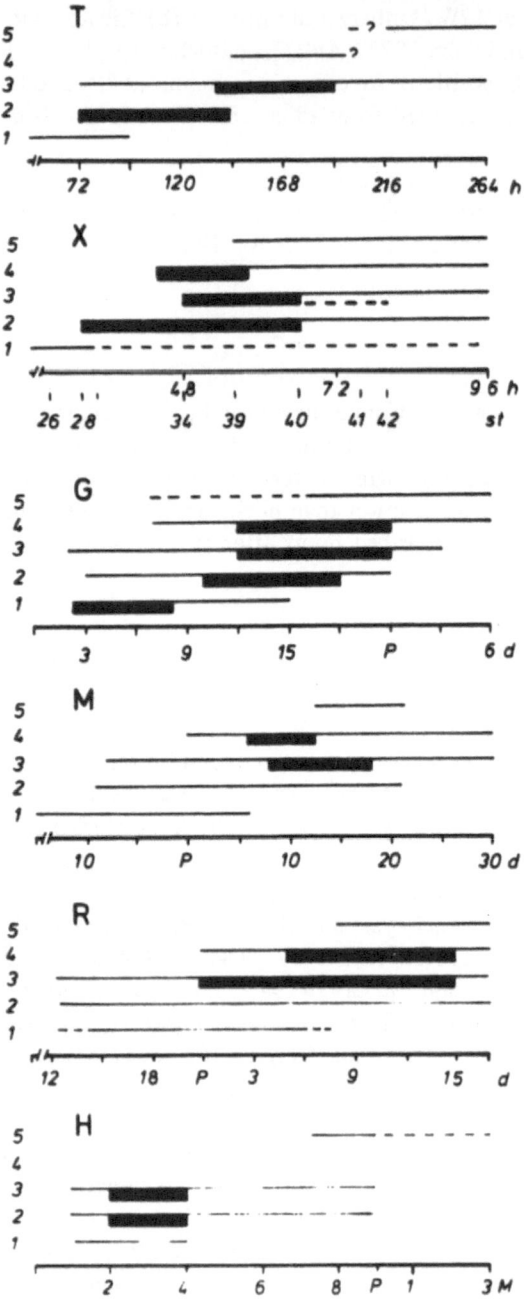

Fig. 4. Temporal order and duration of the various developmental processes in the retina of Tilapia (*T*), Xenopus (*X*), the chick (*G*), the mouse (*M*), the rat (*R*), and man (*H*). Numbers *1* through *5* indicate the developmental phases as defined in the text: *1*, proliferation and migration; *2*, general differentiation including layering; *3*, specific differentiation; *4*, neurochemical differentiation; *5*, specific activity. *Horizontal bars* indicate onset, duration, and presumable end of these processes, with *thick sections* representing main phases

mouse, GABA is the first component. But this may well be a question of the investigations having been or not having been done in this area.

The appearance of specific physiological functions is the last phase to begin. In man, it starts very late, when all other phases have long before been completed. Perhaps this is a consequence of the difficult accessibility of living human embryos before the 8th month; the first ERG might be present earlier.

Neither the switch from embryological to larval stages in *Tilapia* and *Xenopus*, nor the process of metamorphosis in the latter, nor the date of birth in the mammals and in man seem to have any bearing on the development of the retina or the retinal function. In the retina of the precocial bird *Gallus*, however, all developmental processes which occur within the egg shell seem to be adjusted in such a way as to guarantee the ability to see as soon as the chick leaves the egg. This seems to be even more emphasized in the duck and the quail (Heaton 1973; Heaton and Munson 1978). In the rat and the mouse, essential parts of the retina development are extrauterine, thus differing from the guinea pig (cf. Bornschein 1959; Spira 1975, 1976). In the sheep the first visual pigment is found on day 85 of gestation, while the term is after 145 days (Höglung, Nilsson and Schwemer, personal communication). In man, structural differentiation starts very early, whereas retinal activity is found rather late, but both processes are intrauterine. As these differences found in mammalian species are not restricted to retina development, but represent a more profound diversity, they will not be further speculated on here. Apparently, among the species treated in this article, only *Gallus* and *Homo* are capable of receiving light impulses (seeing?) at hatch or birth, respectively. *Xenopus* cannot see when the larva hatches (about st. 38), and *Tilapia* also develops its ability to see only at the larval stage.

2.8 Temporal Sequences

From the appearance of early nineteenth century works till now, a developmental sequence of retina differentiation proceeding from vitreal to scleral zones has been the accepted scheme. It was mainly based on the light microscopic impression of layering and overt cellular differentiation or on the sequence of the last mitotic cycles within the retina (e.g., Babuchin 1863; Weysse and Burgess 1906; Mann 1928; Barber 1955; Hollyfield 1972; Hughes and LaVelle 1974; Kunz and Wise 1974). More recently, however, it was pointed out by electron microscopic or other studies (Olney 1968b; Morest 1970; Kahn 1973a) that this sequence is not always strictly vitreo-sclerad.

The set of data presented here offers an opportunity to test the concept of a vitreofugal sequence on a species as well as on a comparative level.

There can be no doubt about a vitreofugal gradient of layering (Fig. 5). The ganglion cells are invariably the first to begin with the layering, followed by the inner plexiform layer and the amacrine cells, and the bipolar cells are regularly among the last ones. In most cases, however, the outer part of the retina, the receptor cells, and the outer plexiform layer begin their layering before bipolar cells do so. The vitreo-sclerad sequence extends merely to the bipolar cells, not further. There are difficulties in determining the onset of layering and, subsequently, differences among the authors occur. Therefore the notion of vitreo-sclerad development was tested by applying other types of data.

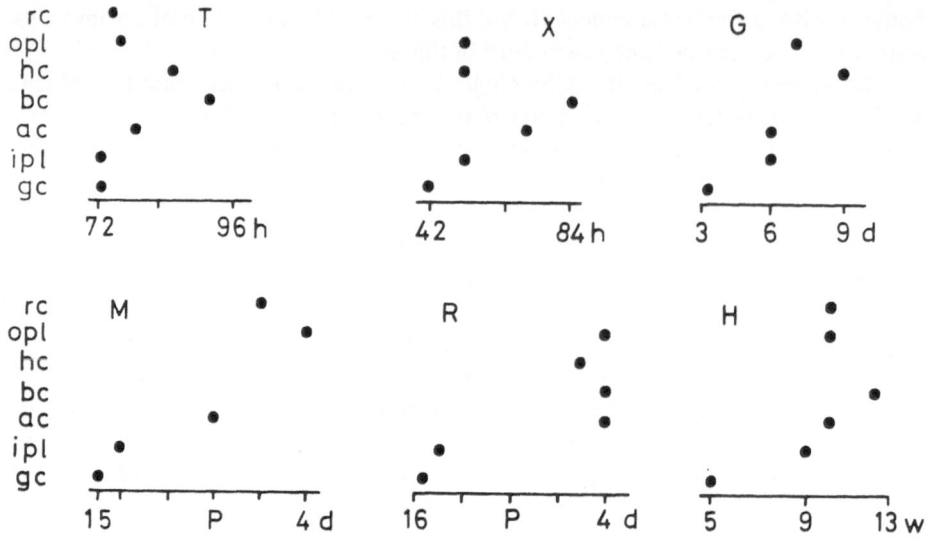

Fig. 5. Time of layering as indicative of light microscopically recognizable onset of differentiation in the six species in question. *Black dots* show first reported onset of layer formation of each of the layers shown at the left border (*rc*, receptor cells; *opl*, outer plexiform layer; *hc*, horizontal cells; *bc*, bipolar cells). A sequence is apparent between ganglion cells and bipolar cells, while outer plexiform layer and receptor cells mostly do not continue this sequence. A vitreo-sclerad temporal gradient extends only up to the bipolar cells

These items, which have been selected according to the functional pathway within the retina, but also according to the availability of pertinent data, are:

Ganglion cell:
　1. Neurites
　2. Dendrites or inner plexiform layer
Inner plexiform layer:
　3. Synapses, conventional or ribbon
　4. Synaptic vesicles
　5. Components of transmitter systems as localized in the inner plexiform layer or
　　　other parts of the retina
Receptor terminal:
　6. Dendritic invaginations (mostly horizontal cell)
　7. Synaptic junctions
　8. Synaptic ribbons
　9. Synaptic vesicles
Receptor cell inner segment:
　10. Ellipsoid, or general aspect
Outer segment:
　11. Disks
　12. Visual pigment
Adult functions:
　13. Electroretinogram
　14. Behavioral or other light responses different from 13.

32

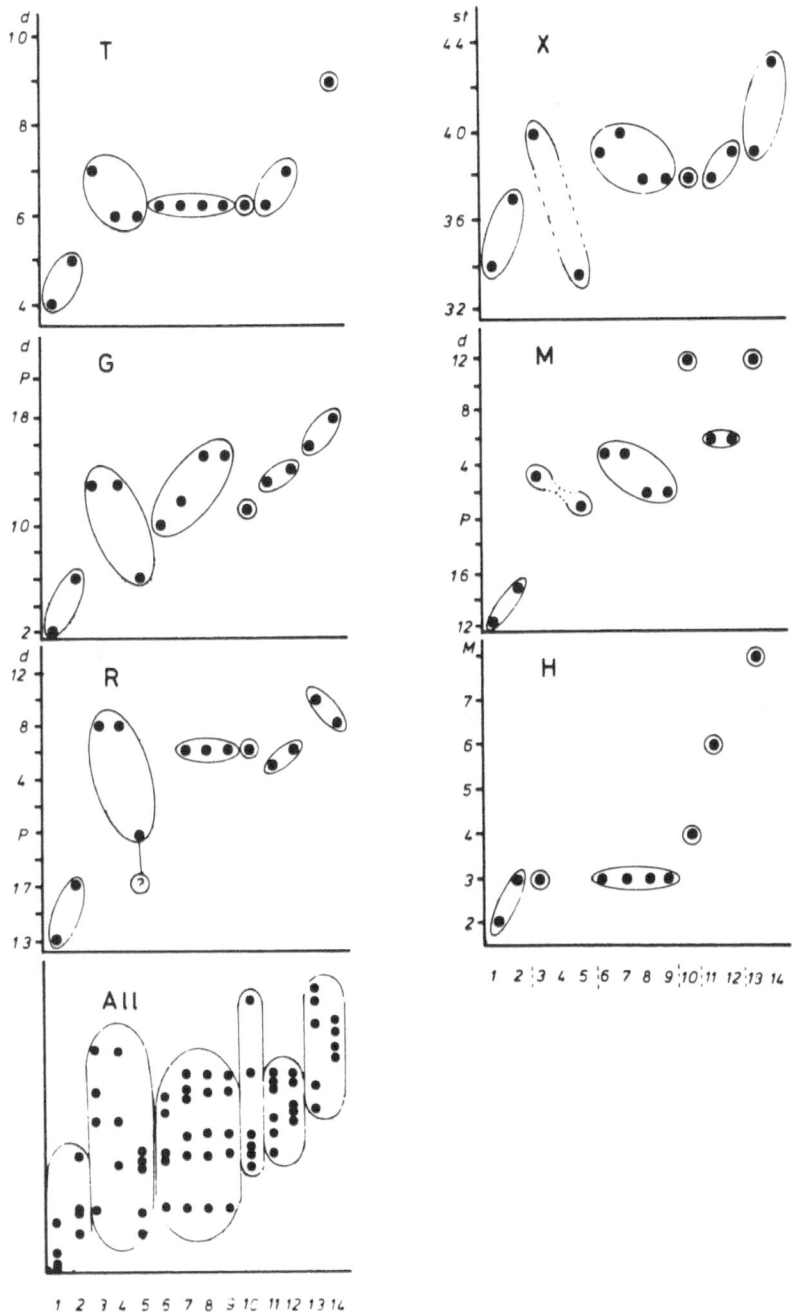

Fig. 6. Time of first appearance of selected retinal structures, substances, and events in the six species as well as averaged for all of them. The various items *1* through *14* as listed on p. *32* are represented by *dots* and arranged (*encircled*) as follows: *1, 2* ganglion cell; *3–5* inner plexiform layer; *6–9* receptor cell terminal; *10* inner segment; *11, 12* outer segment; *13, 14* visual function. For discussion see text

In Fig. 6, the first reliable statement of any of these items is indicated, without regard to the discrepancies between the authors as well as to temporal shifts between different zones of the retina. Again, only the most central zone has been considered.

As stated above, in all cases the ganglion cell neurites (1) are the first structures to appear. (*Xenopus* offers an exception insofar as GABA is present even at an earlier stage, but no precise localization is given, and it certainly does not function as a transmitter at this stage.) Ganglion cell dendrites (2) follow soon, and it may be said that differentiation of the retina in fact starts with the ganglion cells.

No clearly graded development can, however, be found for the functional sites of the inner plexiform layer and the receptor cells (3–12). Apart from differences in details, which cannot be discussed here, these functional sites appear more or less synchronously, leading to the sudden appearance of the physiological maturation of the retina (13, 14).

While the receptor cells mostly differentiate within a short temporal range (6–12), the time span for the inner plexiform layer is broader, due to the relatively early appearance of transmitter substances. But in general there is no great temporal shift between the inner plexiform layer and the receptor cells; in some cases it is even a synchronous development of both. However, the ERG and other responses often evolve after a certain delay, which is probably due to the fact that their appearance – or their registration – needs a larger amount of mature cells.

There are tendencies to a sclerally directed sequence. Ganglion cell and bipolar cell dendrites invariably appear after their axons, and bipolar cell ribbon synapses appear later than amacrine cell conventional synapses. But the concept of a scleropetal advancement of retina differentiation, based on light microscopic observation, has to be replaced for the development of functionally important sites by a three-step model:
1. Differentiation begins in the ganglion cells
2. Simultaneous development in interneuronal and receptor cells
3. Appearance and development of light responses as indicative of a mature retina

From this result it must be concluded that layering and functional differentiation are two processes which are not bound to be interconnected in a direct way.

In the following, the possibility will be tested that not the beginning of differentiation, but the end, the attainment of functionability, reveals a scleropetal gradient.

In order to compare the ages when the first cells achieve functional maturity, a scale is introduced, on which the completion of the eye cup represents zero and the onset of an ERG or a light response is 100. Thus the relative distances of any maturational event can be measured. This has been done in Fig. 7. Apparently, there are clear species differences in the sequence of the maturation of different cell types. (This will become still more pronounced when horizontal cells are included. In all known cases, they attain maturity before bipolar cells do so.) While only *Gallus* and *Homo* fit the old idea of an outwardly directed progress, in *Tilapia, Mus*, and *Rattus* the bipolar cells are the last ones to attain maturity. In *Xenopus*, both types of intermediate neurones are the latest to mature. Again, the only feature common to all species is the fact that the ganglion cell is the first one to attain maturity, as it was the first one to start differentiation, and the first one in layering.

In spite of the differences, these data were averaged, and, for a generalized vertebrate scheme, a temporal gradient between the inner and the outer half of the retina results, but again, the receptor cells are more advanced than the bipolar cells.

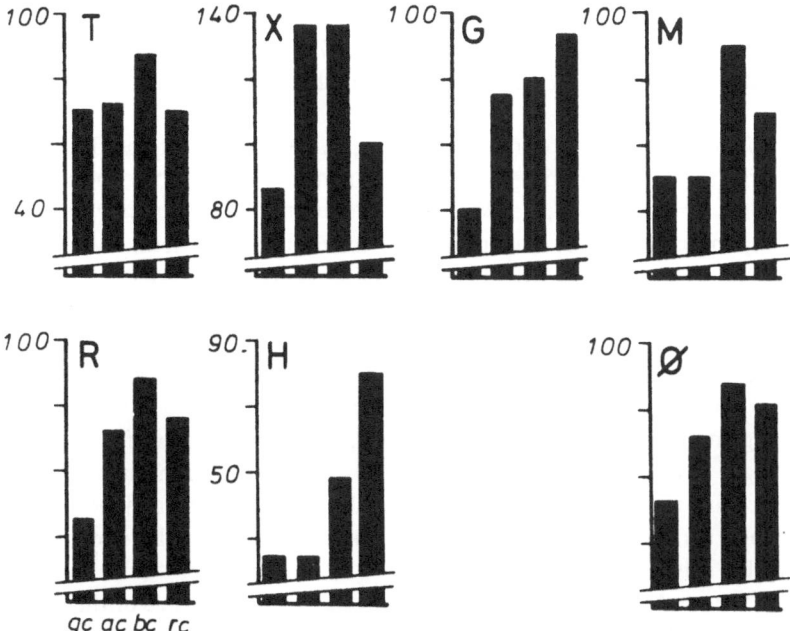

Fig. 7. Relative time of maturation in the ganglion cells (*gc*), amacrine cells (*ac*), bipolar cells (*bc*), and receptor cells (*rc*) in each of the six species and as averaged (*∅*). On the ordinate scale, 0 is the point of eye cup formation and 100 represents first appearance of an ERG or light response. Again there is a sequential gradient between ganglion cells and bipolar cells, the receptor cells being mostly more advanced than the latter

To summarize, by regarding the attainment of function, we also come to the conclusion that the concept of an outwardly directed differentiation must be abandoned in favor of a development in clusters, ganglion cells — intermediate neurones — receptor cells. Physiological maturity is achieved neither gradually, in a unidirectional sequence, nor synchronously. This result leads to the suspicion that the development of function in the tissue "retina" is not so much a matter of coordination and regulation, but is rather brought about by a program inherent to the single cells.

3 Specific Differentiation

3.1 Receptor Cell Inner Segment Formation

The receptor cell inner segment is essential for the metabolic and energetic requirements of the specific part of the receptor cell (Rodieck 1973). Furthermore, there are specific components — an "oil droplet," subsurface cisterns, and others — which suggest a specific participation in the processes of reception.

In the following, the development of the inner segment will be summarized for the teleosts *Haplochromis* (Rohrschneider 1975), *Tilapia* (Grün 1975, and unpublished results), and *Poecilia reticulata* (the guppy; Berger 1964; Wise and Kunz 1977); for *Ambystoma* (Detwiler and Laurens 1921) and other amphibians (Carasso 1958); the chick (Meller and Breipohl 1965; Meller 1968; Olson 1972, 1973, 1979; Meller and

Tetzlaff 1976; Mishima and Fujita 1978); the mouse (De Robertis 1956; Feeney 1973); the rat (Bok 1968; Feeney 1973; Galbavy and Olson 1978); the dog (Hebel 1971); the rhesus (Smelser et al. 1974); and *Homo* (Yamada and Ishikawa 1965; Hollenberg and Spira 1973).

At first the inner segments appear as spherical apical extrusions of the future receptor cells, characterized by ribosomes, polysomes, microtubuli, and some mitochondria (Fig. 8). Some cisternae of the endoplasmic reticulum, usually of the granular type, are present, and a Golgi apparatus is formed, scleral to the nucleus. Over a certain period of time, the amount of ER and ribosomes increases, but most conspicuous is the increase in the number of mitochondria. They are orientated, more or less parallel to the long axis of the receptor cell, and accumulate in the apical region. In the guppy, thereby a sclero-vitreally directed gradient in size and complexity arises. In the final stage, a true ellipsoid has been formed, consisting of densely packed mitochondria.

With the beginning of inner segment formation, two centrioles are found in its apical part, arranged at right angles. Later, one of them, which is to form the outer segment, produces ciliary rootlets. At this time, microvilli have appeared and increase in length and number, surrounding the apical part of the inner segment. Later on, some of them become the calycal processes.

In the chick, the paraboloid takes its origin from vesicles and cisternae of the endoplasmic reticulum, which then form a network of increasing extent and complication. Only in the latest stages do glycogen granules appear in and between the vesicles. Ac-

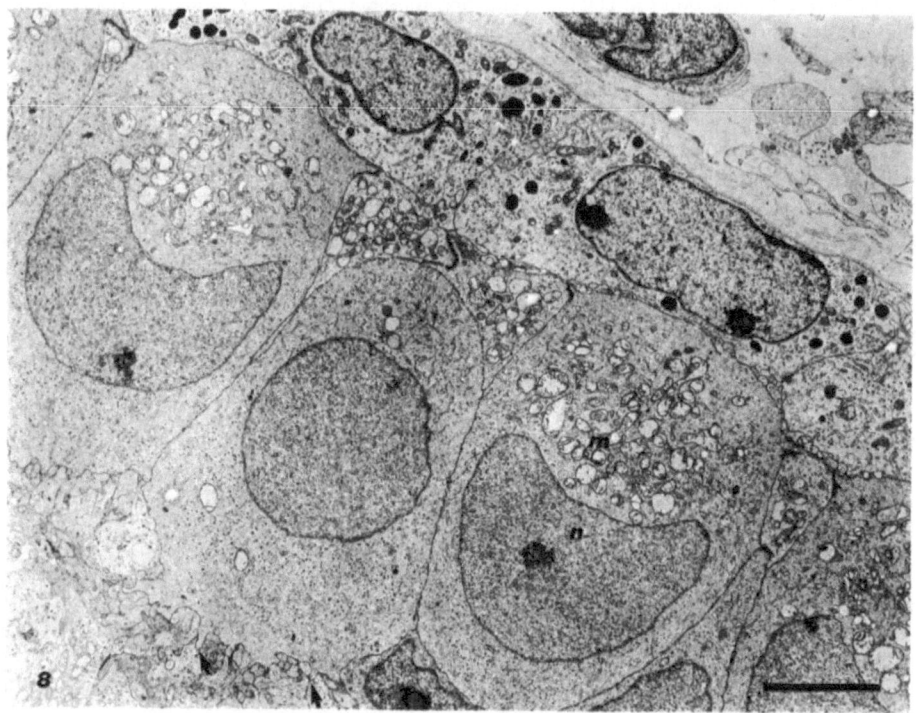

Fig. 8. Developing cones in a 76-day gestation retina of *Macaca mulatta*. Ribosomes and ER are characteristic of cells of this stage. The cytoplasm apical to the large nucleus (*n*) is progressively filled with mitochondria (*m*), while the basal cytoplasm protrudes between processes of the outer plexiform layer (*arrows*). Bar = 10 µm. (Smelser et al. 1974)

cording to Carasso (1960), in the urodelan *Pleurodeles*, the ER profiles disappear when glycogen particles accumulate.

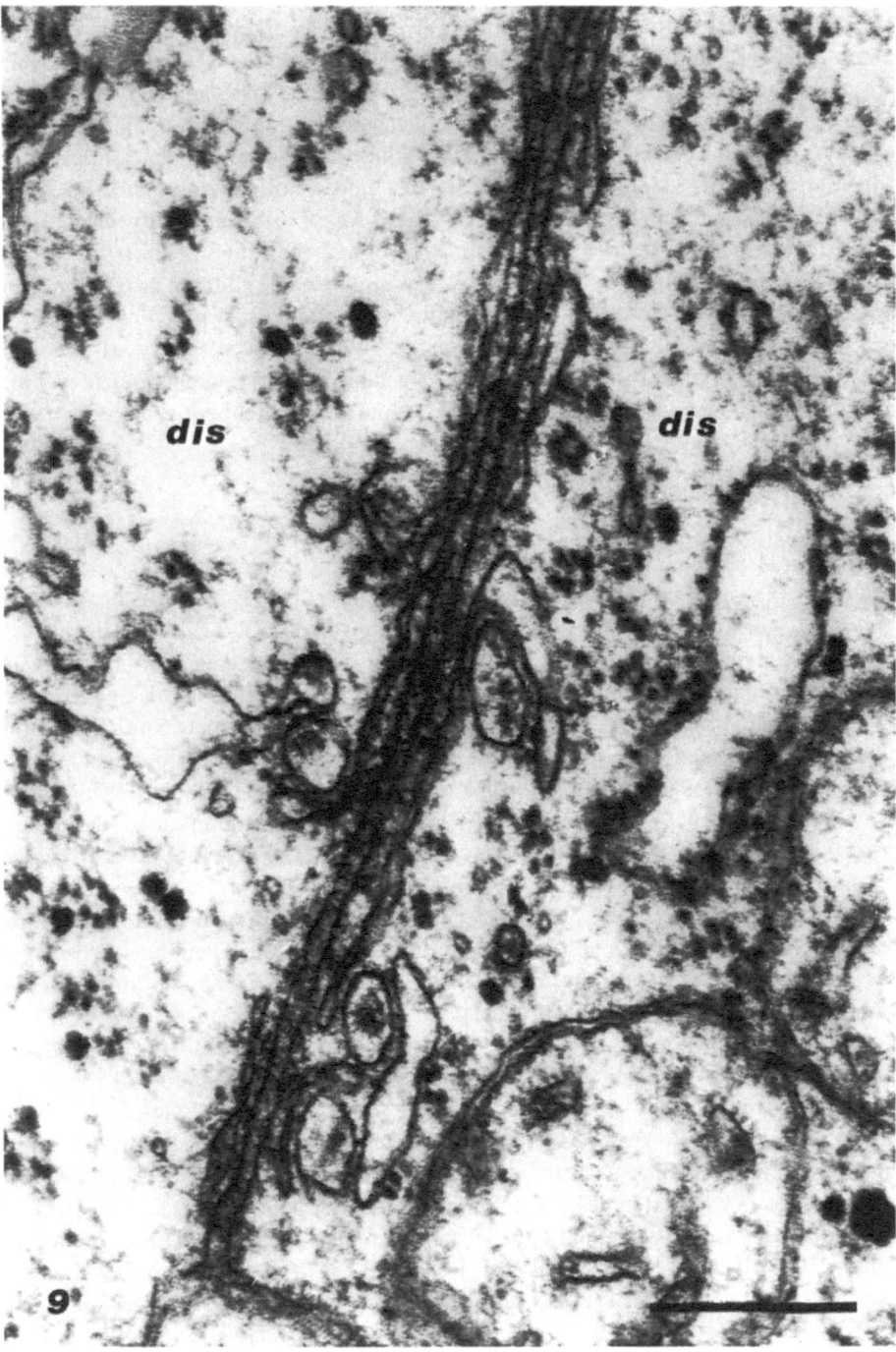

Fig. 9. Formation of subsurface cisterns in double cone inner segments (*dis*) of *Tilapia leucosticta* by fusion of flattening ER vesicles. Bar = 0.5 μm

The development of the so-called oil droplet of cone or double cone inner segments has been studied in the chick and amphibia. Coulombre (1955) described the
fusion of many smaller fatty or lipid droplets — the origin of which is not known — as
one big oil droplet in the chick. This explains the observation by Cooper and Meyer
(1968) that inner segment pigment material may spectroscopically be detected long
before being visible. A yellow carotenoid, galloxanthin, and the red astaxanthin, which
are characteristic of the double cone accessory element, the main element, and the
single cones, respectively, appear in this sequence on different days. Oil droplets are
usually confined to cones, but in the frog, *Rana*, they occur in some future rods,
which are therefore assumed to pass a cone-like stage (Yoshida and Ninomiya 1967).
Cone-like stages in rods have also been suggested for *Ambystoma* (Detwiler and Laurens 1921), *Xenopus* (Kinney and Fisher 1978b), and the rat (Brockhoff 1957). Subsurface cisterns, which are typical of teleostean double cones, but are also found in adjacent cones of the human eye (Linberg and Fisher 1978), are formed by a synchronous fusion and flattening of vesicles. This process of fusion occurs very early, coexten-
sive to the region where the two elements of a double or a twin cone touch very inti-
mately (Fig. 9) (Berger 1967; Ahlbert 1973; Grün 1977a; Linberg and Fisher 1978).

3.2 Receptor Cell Outer Segment Formation

In the first electron microscopic study of outer segment formation De Robertis (1956)
described the outer segment of the mouse retina as taking its origin from tubuli and
vesicles which he regarded as the "morphogenetic material" for disks. This material
was said to derive from the cilium growing out of the inner segment and by flattening
and elongation to lead to the formation of disks. Later (De Robertis and Lasansky
1961) he stated that the "subsurface membrane actively participates in the formation
of primitive rod sacs." The formation of outer segment disks from tubules and vesicles
was subsequently reported for the chick (Meller 1964, 1968), the rat (Weidman and
Kuwabara 1968, 1969), the dog (Hebel 1971), the cow (Mason et al. 1973), and perhaps the rhesus (Keefe et al. 1966), as well as for the regeneration of disks after outer
segment loss (Dowling and Gibbons 1961).

As early as 1959, however, Lanzavecchia described the disk formation in *Xenopus*
as being the result of unidirectional invagination of the plasma membrane. Working on
several amphibian species, Carasso (1958, 1959) described membrane infoldings, but
stated that disks are formed out of vesicles.

According to the careful study of Nilsson (1964) in the leopard frog, *Rana pipiens*,
the first indication of disk formation is a funnel-shaped invagination of the plasma
membrane around the ciliary connecting piece. While being sclerally displaced, this invagination enlarges and gives rise to a flattened saccule, a disk. Vitreal to it, new invaginations and new saccules are formed. No signs of vesicles were found. Similar ways of
disk formation were later described for man (Yamada and Ishikawa 1965), the rat
(Bok 1968; in vitro: LaVail and Hild 1971), the mouse (Olney 1968a), the guinea pig
(Spira 1975), the rabbit (McArdle et al. 1977), the cat (Vogel 1978a, b), several other
mammals (Anderson et al. 1978), the chick (Govardovsky and Kharkeevich 1965,
1966). *Xenopus* (Kinney and Fisher 1978b), the teleosts *Tilapia* (Grün 1974, Fig. 6),
and *Poecilia reticulata* (Wise and Kunz 1977), the ammocoetes larva of an unknown
cyclostome species (Vinnikov 1969), and even for the larval eye of the ascidian *Ama-*

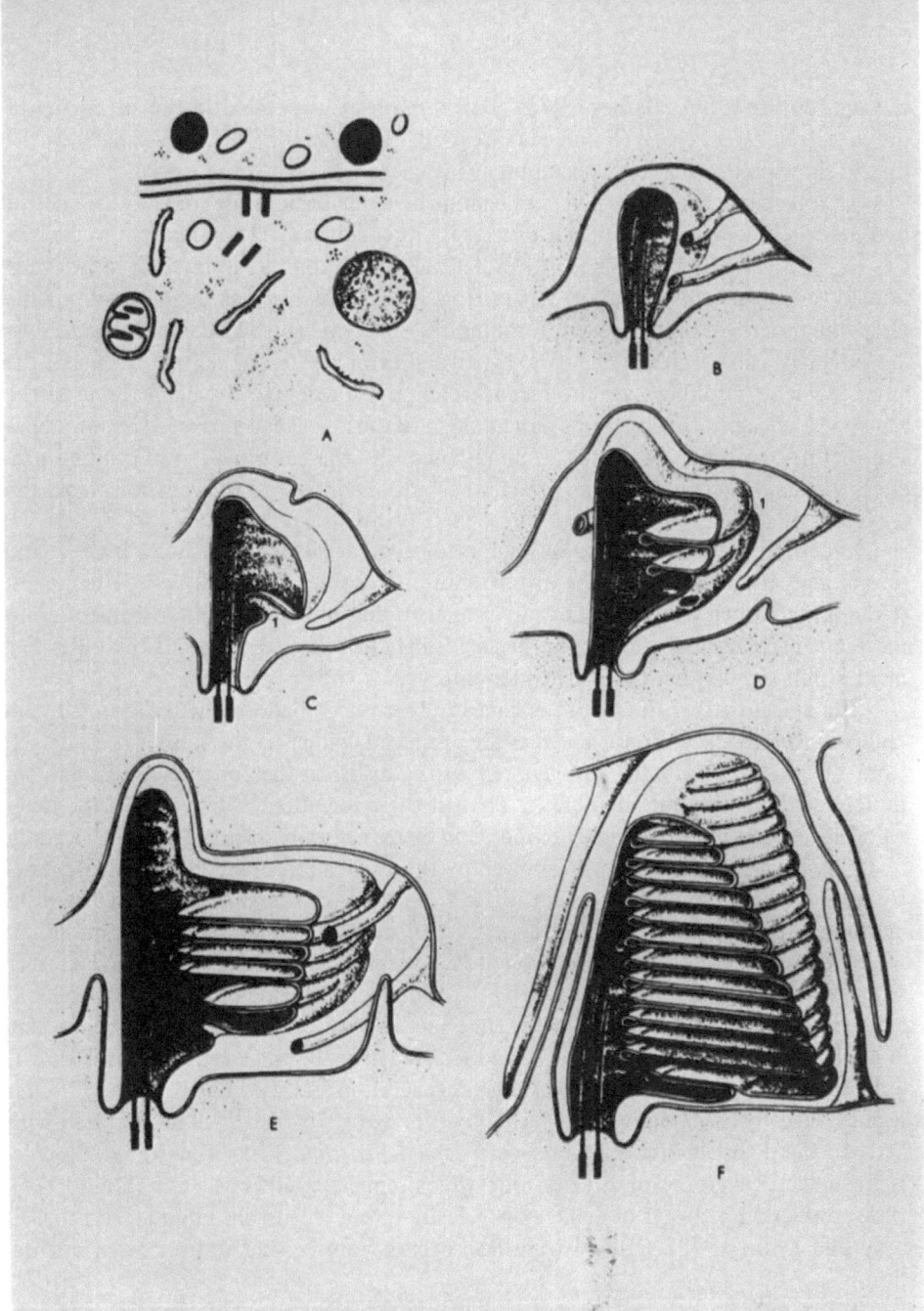

Fig. 10. Schematic illustration of the process of outer segment formation in the tadpole retina (*Rana pipiens*). A centriole approaches the inner segment plasma membrane (*A*) and a ciliary stalk is produced (*B*). The expanding ciliary stalk is invaginated (*C*) and shifted sclerally, thereby enlarging to form a disk. This process is repeated (*D–F*). (Nilsson 1964)

39

roucium constellatum (Barnes 1974). Furthermore it was found in the stirnorgan of the tree frog *Hyla regilla* (Eakin and Westfall 1961), and in fact, this paper was the first to show disk formation by membrane invagination.

As apparently species differences cannot be made responsible for the existence of two models of disk formation, this process is either subject to variation within the species, or a choice has to be made. At present all indications are in favor of disk formation by membrane invagination. Apart from the discrepancy as to the origin of the disks, the process of outer segment formation is quite well agreed upon, and the following summary can be given (Fig. 10): On two sites of the apical cell membrane of the inner segment, infoldings are produced, which are directed to the nearest one of the centrioles, which normally forms a right angle to the apical cell surface. The infoldings stop before reaching the centriole. Simultaneously, the centriole grows out and produces a ciliary stalk, which is a true cilium, but, in contrast to the centriole, lacks the two central tubuli.

In certain teleosts, but also in man, a so-called accessory outer segment is found (Munk and Anderson 1962), the ultrastructure (Yacob et al. 1977) and development (Grün, unpublished work) of which are similar to that of a normal outer segment up to the moment when disk production begins. It may be said that the accessory outer segment differs only in that disks fail to develop.

The apical part of the cilium enlarges and assumes a balloon shape. Then — by invagination or vesicle fusion — saccules are produced, which, in the case of rods, detach from the plasma membrane and give rise to the disks. In the cones they remain continuous with the plasma membrane. Though these membranes derive from the inner segment, a change in membrane composition seems to occur, as is suggested by a study of Bridges and Fong (1979), who showed by the application of lectins that membranes from the base of the outer segment differ from those of the inner segment. The disks increase in size by fusion or rearrangement of membrane material among infoldings which are not yet detached. Synchronously, the disks are shifted sclerally and another series is produced. According to Ditto (1975) disks in the cone outer segments of *Ambystoma* are produced simultaneously, not successively, and the same may be the case in adult and larval *Xenopus* (Kinney and Fisher 1978a, b, c). Questions of this kind as well as the closely related problem of disk regeneration and shedding in the adult outer segment have been extensively reviewed by Young (1976) and shall not be dealt with here. In the double cones of the teleost *Tilapia leucosticta*, disks were found to be detached in vertical groups of three or four, which remain together as a unit (Grün 1974). Disk production is always directed excentrically to one side of the cilium (see also Galbavy and Olson 1978). Often a transition occurs from vertical to horizontal orientation.

A few differences other than those already mentioned are observed between rods and cones in the earliest stages.

The rapid increase in length following the first appearance of disks is accompanied by an increase in volume, which, in *Xenopus*, has been calculated to occur at a constant rate for at least 2 weeks (Kinney and Fisher 1978b). Final length and velocity of growth, however, are regulated by a varying relation in the rates of production and shedding (LaVail 1973).

Parallel to this increase in length and volume an increase in visual pigment content is noted. For *Xenopus* Witkovsky et al. (1976) was able to show an increase in the total rod outer segment volume and in the amount of visual pigment molecules in the

40

whole eye between st. 39 and st. 58. Nevertheless the visual pigment concentration remained constant. The number of molecules/μm^3 outer segment amounted to 2.28 × 10^6 at st. 39 and 2.29 × 10^6 at st. 58, thus clearly showing a parallelism between membrane addition and visual pigment incorporation. Retinol has its highest concentration in the pigmented epithelium and occurs as free alcohol. According to Wiggert and Chader (1975) it binds to a limited number of binding sites on receptor molecules, which are present long before outer segments appear. On the other hand the opsin has been reported to be of inner segmental origin in the frog by Papermaster et al. (1975). It is a membrane-bound molecule which is transported to the apical part of the cell and incorporated into the plasma membrane. Here retinol is apparently bound to a lysine residue of the opsin (Bownds 1967). Bok et al. (1977) describe the rhodopsin biosynthesis as beginning at inner segment ribosomes where opsin starts to migrate to the outer segment. During this migration it is glycosylated. At the base of the outer segment it is conjugated with retinol and incorporated into newly forming disk membranes.

By application of the freeze-fracture technique in the mouse, Olive and Recouvreur (1977) showed that clusters of particles, thought to represent visual pigment, are incorporated into already existing disks (Fig. 11). In adult rods, however, these particles were found in definite number and distribution in newly formed disks, too. That disks might exist without visual pigment as an essential component is also suggested by an observation of Witkovsky et al. (1976). Vitamin A (retinol) deficiency during larval stages of tadpoles of vitamin A deficient mothers led to a visual pigment reduction by 75% while the structural integrity of the outer segment was unaffected. This might be explained in the way that opsin may be incorporated into the disk membrane, but due to the lack of retinol no functioning visual pigment is formed. There are, however, reports showing the importance of retinol for the existence of disks: In the mouse (Dowling and Gibbons 1961) and the lizard *Sceloporus* (Eakin 1964), lack of vitamin A causes degeneration of the disks, at least partially, having no other effects. This may be comparable to the observation that in the absence of a pigmented epithelium only a few, abortive disks are produced (Hollyfield and Witkovsky 1974). Silverstein et al. (1971) even state that most cases of retinal dysplasia occur due to the absence of the normal contact to the adjacent normal pigment epithelium. A correlation between retinal dysplasia and abnormal retina-pigment epithelium relations is reported for man by Fulton et al. (1978).

An interesting ecological and phylogenetical circumstance is presented by the fact that in some fishes and amphibians, larvae show the same preponderance of porphyropsins as freshwater teleosts do, while in postmetamorphic animals only rhodopsin is found, as it is the case in terrestrial vertebrates. In the bullfrog (*Rana catesbeiana*), this change is attained by the mere fact of metamorphosis, as could be shown by stimulation of premature metamorphosis (Wilt 1959). A summarizing review on these and related phenomena is given by Bridges (1972).

The influence of light is not necessary for the complete receptor differentiation in *Hyla regilla* (Eakin 1965) and *Tilapia leucosticta* (Grün 1979) and can also be denied for phagosome production in fetal guinea pigs (Spira and Huang 1978). Hollyfield and Rayborn (1979), however, presented evidence that in *Xenopus*, outer segment development occurs faster in continuous light than in a light-dark cycle, and here faster than in continuous darkness.

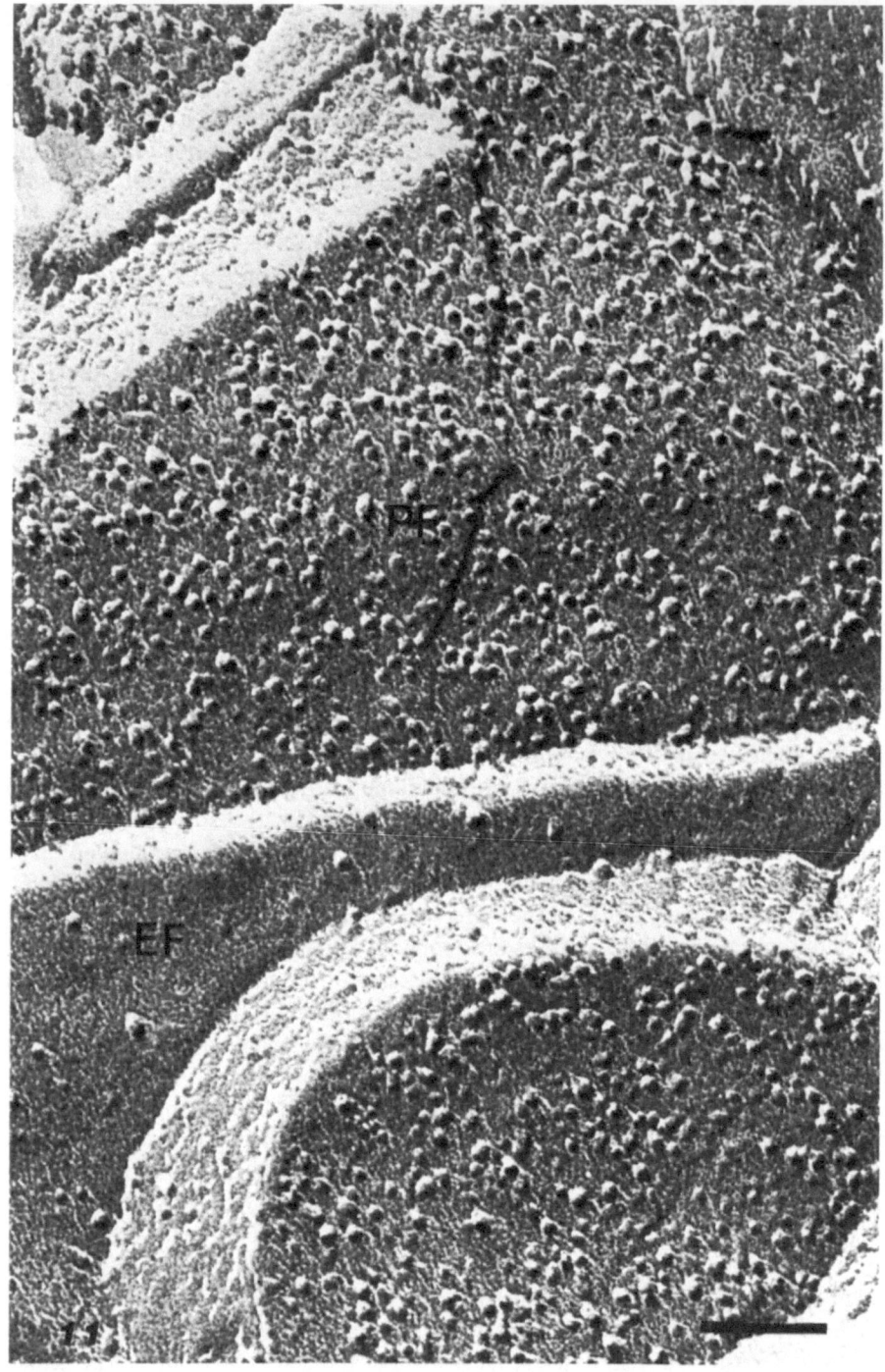

Fig. 11. Freeze fracture preparation of mouse outer segment, d8pp. Alternating fracture faces (*EF*, *PF*) are exposed. Intramembraneous particles thought to represent visual pigment are already found at one fracture face (*PF*), but their number is low as compared to later stages. Bar = 0.1 μm. (Olive and Recouvreur 1977)

3.3 Receptor Cell Terminal Formation

At the onset of its general development, the receptor cell terminal is a small cytoplasmic stripe, vitreal to the nucleus. It mainly contains ribosomes, either free or arranged as polysomes. Some profiles or vesicles of the ER may occur, as well as dendritic growth cones. Microtubuli are mentioned by some authors only. With the proceeding development, the number of ribosomes increases, but later it is distinctly reduced. These general observations apply to the cichlid *Tilapia* (Grün 1975), *Xenopus* (Grün, unpublished work), the chick (Meller 1968; McLaughlin 1976a), the mouse (Olney 1968a, b), the rhesus monkey (Smelser et al. 1974), and man (Yamada and Ishikawa 1965; Hollenberg and Spira 1973; Spira and Hollenberg 1973). In the course of the further development, the terminal region steadily increases in volume.

In some cases (chick: Meller 1964; rat: Weidman and Kuwabara 1969; cat: Vogel 1978a; rhesus: Smelser et al. 1974) this process seems to be preceded by an outgrowth of cytoplasmic processes from the basal cell surface. The number and branching of these processes increase while they penetrate, more or less deeply, the now originating outer plexiform layer where they intermingle with processes from the intermediate neurones.

The latter ones contact the processes or the broad terminal surface. In either case, subsequent depressions of, and invaginations into, the receptor terminal occur. It seems to be a general rule that the first two horizontal processes form a dyadic arrangement, postsynaptic to a terminal synapse. It is only later that a bipolar cell dendritic process enters as a central element, thus yielding a triadic configuration (*Tilapia*: Grün 1980; *Xenopus*: Chen and Witkovsky 1978; *Rana catesbeiana*: Nilsson and Crescitelli 1969; chick: Meller 1964; McLaughlin 1976a; mouse: Blanks et al. 1974a; rabbit: McArdle et al. 1977; cat: Vogel 1978a; rhesus: Smelser et al. 1974; man: Linberg and Fisher 1978). For *Xenopus* (Chen and Witkovsky 1978), the rabbit (McArdle et al. 1977), and the human retina (Hollenberg and Spira 1973; Spira and Hollenberg 1973) the existence of a single postsynaptic process during early stages has been reported. In *Xenopus*, changing relationships between the number of dendritic invaginations and synaptic ribbons were found.

As a consequence of invagination, synaptic sites are produced along the more and more extending basal receptor cell surface. In the chick (McLaughlin 1976a,b), pre- and postsynaptic membranes are situated in a parallel order, and the former is thickened. The characteristic arciform density appears at those points where the two or three postsynaptic processes touch the presynaptic membrane. Synaptic cleft densities and postsynaptic dense projections, as shown by the ethanolic-phosphotungstic acid technique, arise in the following. Later on the stainability is enhanced. In the course of receptor terminal development, the lectin concanavalin A (McLaughlin and Wood 1977) as well as the wheat germ agglutinin and ricinus agglutinin (McLaughlin et al. 1980) bind to synaptic membranes, but also to the undifferentiated receptor cell membrane. After hatching, when adult structure of the receptor terminals can be assumed, only nonsynaptic membranes bind the lectins. Thus differentiation of synaptic sites is supposed to lead to a masking of corresponding binding sites. Martinozzi and Moscona (1975), however, describing similar phenomena in dissociated cells of the chick retina, suggested a decrease of the lateral mobility of lectin receptors. The occurrence of binding sites at nondifferentiated membranes only might be correlated to a role in the process of differentiation or in the process of specific recognition.

In the same species, Meller and Tetzlaff (1977) stated by the use of the freeze-fracture technique that presynaptic membranes show increasing amounts of synapto-pores which may represent synaptic vesicles. On dendritic outer membrane surfaces aggregations of particles were found which form a lattice structure after some time. These particles may be interpreted as transmitter receptors.

In nearly all of the hitherto mentioned species, synaptic vesicles, when formed for the first time, are low in number and occur either freely distributed or as a vesicle halo around a synaptic ribbon.

Origin and development of the synaptic lamellae or synaptic ribbons, which probably serve as a kind of guiding structure for synaptic vesicles, are most controversial. The first synaptic lamellae are either found in paranuclear regions (Meller 1964, 1968; Olney 1968a; Smelser et al. 1974), or floating freely in the cytoplasm (Vogel 1978a), or at synaptic sites (McLaughlin 1976b; Grün 1980). This has given rise to the assumption that they may take origin near to the nucleus, far from their future sites, to which they migrate singly or in clusters later on. In the guinea pig, Spira found the first synaptic ribbons either in a paranuclear region, or in numbers of up to four at future synaptic sites; in the rabbit, McArdle et al. (1977) observed synaptic lamellae in paranuclear regions, or floating freely in the cytoplasm, or rarely in contact with one or more postsynaptic processes; in man, Linberg and Fisher (1978) localized them at synaptic sites or free in the cytoplasm. These latter observations shed doubt on any theory of synaptic lamellae origin, unless one assumes the existence of different modes of forma-

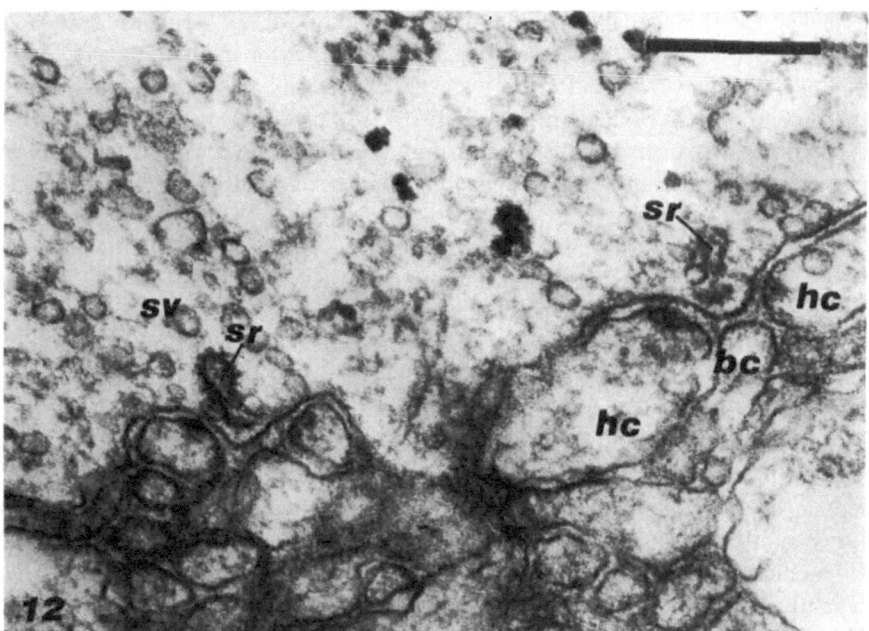

Fig. 12. Six-day-old *Tilapia leucosticta*. Receptor cell terminal with synaptic vesicles (*sv*) and two examples of synaptic ribbons at the very beginning of their formation (*sr*). When these ribbons appear, triadic configuration of the dendritic invagination is already present. *hc*, horizontal; *bc*, bipolar cell dendrite. Bar = 1 μm

tion in different receptors. In *Tilapia*, Grün (1980) found the very first lamellae – invariably the smallest ones – already and exclusively in their definite positions (Fig. 12); only during later stages were they occasionally found to be localized all over the terminal up to paranuclear regions. He also documented a growth of these ribbons, thus confirming other authors (e.g., Chen and Witkovsky 1978).

In the mouse (Olney 1968a), the rabbit (McArdle et al. 1977), and man (Hollenberg and Spira 1973), the very first ribbons are already surrounded by a halo of synaptic vesicles as it is characteristic of adult stages. Weidman and Kuwabara (1968, 1969) described electron-dense granules in the rat retina, which arise among synaptic vesicles and aggregate into synaptic ribbons. Similar illustrations can be found in Shiragami's study (1968) on the chick retina. Even amorphous formations or dense bodies made up of granulations and surrounded by a Golgi vesicle membrane have been regarded as ribbon precursors (Smelser et al. 1974). These structures, however, bear no resemblance to a synaptic ribbon into which they have to be suddenly transformed.

At present, no unifying idea on synaptic ribbon formation can be offered, the less as Spadaro et al. (1978) mention a further mode through which synaptic lamellae may be formed by infolding and detachment of the receptor cell terminal membrane. It is assumed here that the picture is further obscured by the fact that not in every case have the very earliest ribbons or ribbon precursors been described.

As far as the temporal sequence by which the functional elements of the receptor cell terminal appear is concerned, there seem to be no regularities. In most cases differences, however, are due to incomplete data. The most comprehensive set of data is available for the chick. Summarizing all data and neglecting minor differences, one can establish the following sequence: Contacts formed between the receptor terminal membrane and processes of the intermediate neurons are followed by the invagination of the latter into the terminal. At first, horizontal cell invaginations form a dyad onto a synaptic region; later, a bipolar cell process penetrates, central to these, and a triad is formed. At the sites of contact, membrane thickenings now mark the position of synapses which, in the following, are further characterized by cleft densities and postsynaptic densities. The last structures to appear are synaptic vesicles and synaptic lamellae. This sequence strictly follows a path from outside the terminal to intracellular events, suggesting not only a temporal but also a causal sequence.

Another type of sequence is found in rodents (*Mus, Rattus, Cavia*), primates (*Homo, Macaca*) and, with certain deviations, also in the frog (*Rana pip.* and *R. cat.*) and in the fish *Tilapia*. Again, invaginations are the first features more or less immediately followed by the appearance of synaptic lamellae and synaptic vesicles. It is only then that membrane thickenings and cleft densities are found. The arrangement of invaginations in dyadic and then triadic form has occasionally been reported to occur as the last event. In *Tilapia* it is finished before synaptic ribbons are produced.

The rabbit, the cat, but also *Xenopus laevis* do not start the receptor terminal differentiation with an invagination of neuronal processes. Here, the ribbons seem to be the first structures. In *Xenopus*, basal contacts precede ribbon formation, but it is not certain whether these are part of the differentiational process or only represent a consequence of close topographical relationship. Ribbon formation in *Xenopus* is accompanied by a halo of synaptic vesicles. Invaginations are formed later on, and the maturation of the synaptic membranes is the last step. In the rabbit and the cat, synaptic sites appear after or simultaneous to ribbons. Dyadic invagination may also be formed at that stage. Triads are preceded by synaptic vesicles.

No doubt, the data are incomplete and do not yet allow final conclusions. The "first" appearance of, say, synaptic vesicles is easily overlooked. Perhaps further observations will establish a clear temporal and causal correlation. At present it can be said that the structural components of the synaptic junctions (membrane thickening, etc.) mature in close temporal correlation and that, with exception of the frog and two mammalian species, synaptic ribbons and synaptic vesicles always appear simultaneously.

3.4 Formation of Functional Sites in the Inner Plexiform Layer

The outgrowing neuronal processes separate the hitherto adjacent amacrine from the ganglion cells; (there is no study on the conditions of specification and specific connectivity in a neuronal network, which is brought about not by search and encounter of formerly separated cell processes, but by a separation of formerly adjacent ones.) Initially, they are loosely packed and give rise to a zone with numerous intercellular spaces. These disappear by and by, when more neuronal, but also many Müller cell processes invade, producing a very tight package of the rising layer. This has been observed in a teleostean (Grün 1977b) and several mammalian retinae (Weidman and Kuwabara 1968; Vogel 1978b; Smelser et al. 1973, 1974). An enhanced increase of neuronal processes in the *Xenopus* retina is visualized by a very strong incorporation in the inner plexiform layer during developmental stages of ^{3}H-glucosamine, which is supposed to be a cell membrane precursor (Hollyfield et al. 1975). While this process is taking place, there is a sudden binding of the lectin concanavalin A to the inner plexiform layer (Ulshafer and Clavert 1979), suggesting a change in the configuration and composition of neuronal membranes.

Early outgrowths stem from amacrine or ganglion cells, as had been shown for the chick retina (Shen et al. 1956); bipolar cell processes can be discerned by their specific terminals only later on (Olney 1968b; Grün 1977b).

In the primate retina, the layer of Chievitz is formed, which may be regarded as a precursor of the inner plexiform layer. Smelser et al. (1973) found in it typical neuronal processes containing neurotubuli, neurofilaments, ribosomes, and vesicles. The appearance of junctional and glial processes marks the transition to the final inner plexiform layer, whose processes are supposed to be formed anew. For the human retina it was suggested (Rhodes 1979) that this transition is not a result of intracellular changes, but is merely a consequence of the emigration of ganglion cells from the Chievitz layer. Ganglion cell dendrites and amacrine cell processes meet at various levels of the inner plexiform layer. Even under the light microscope, these levels are recognized as bands, running through the layer, paralleled to the retina inner and outer surfaces. The bands are constituted of synaptic junctions. Other bands are found, made up of cytochemically localized acetylcholinesterase (Shen et al. 1956). Later, synaptic bands of bipolar cell junctions are found.

The formation of synaptic junctions (see Fig. 13) is indicated by the appearance of an asymmetrical thickening, cleft densities, and finally of 40- to 50-nm vesicles close to the junction. This picture has been found to be characteristic of the developing inner plexiform layer in the E-PTA treated retina (Grün 1977b). Often, a transition from desmosomic to synaptic junctions has been assumed (Spira 1975; Smelser et al. 1974; Spira and Hollenberg 1973), but has never been proved. In the rat (Weidman and Ku-

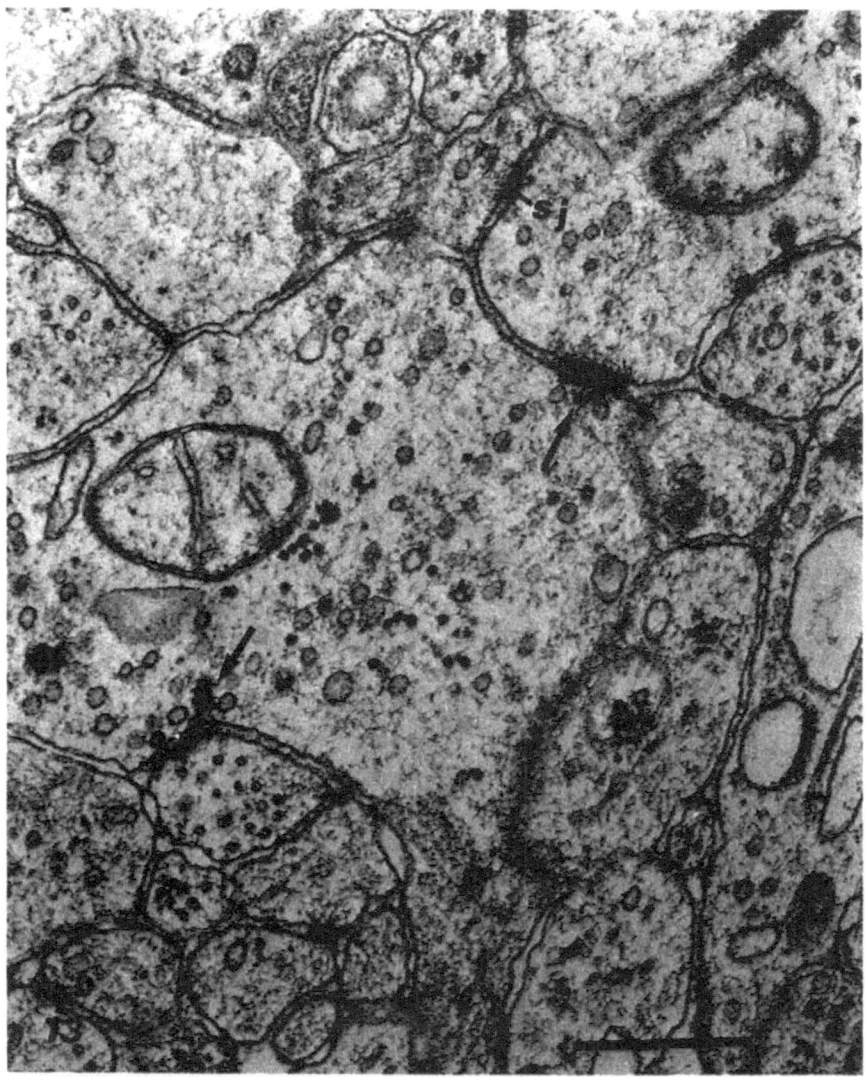

Fig. 13. Inner plexiform layer of a 76-day-gestation rhesus monkey. A bipolar cell terminal with developing synaptic ribbons (*arrows*) and an immature synaptic junction (*sj*) are seen. Bar = 0.5 μm. (Smelser et al. 1974)

wabara 1969; Spira 1974), the guinea pig (Spira 1975), the rhesus monkey (Smelser et al. 1974), and man (Spira and Hollenberg 1973) synaptic junctions are formed long before synaptic vesicles appear in adjacent terminals; in *Tilapia* (Grün 1977b), the mouse (Olney 1968a), and the rabbit (McArdle et al. 1977) both structures are simultaneously formed.

Whether a maturing junction will be a conventional or a ribbon junction seems to be determined by whether it is formed by an amacrine or by a bipolar cell (Dowling and Boycott 1967). Consequently, ribbon synapses appear after conventional synapses have been formed (Spira and Hollenberg 1973; Spira 1974; Cragg 1975; Fisher 1979; McArdle et al. 1977; Vogel 1978a). In the mature retina, a ribbon is presynaptic to

47

two or more postsynaptic processes, which may be of amacrine or ganglion cell origin. In the developing retina of the rabbit (McArdle et al. 1977) and of man (Spira and Hollenberg 1973) stages have been observed where only one postsynaptic process is present. In the rabbit it has been observed that, when the second process penetrates up to the synaptic site, the membrane thickening spreads from the first to the additional contact region. When a synaptic site is present — either dyadic or monadic — a synaptic ribbon appears, and it is only then that vesicles aggregate at the junction (e.g., Weidman and Kuwabara 1968; McArdle et al. 1977; Grün 1977b). It can therefore be assumed that ribbon synapses are in fact functional only when ribbons are present, but not before.

The development of the number of synapses has been studied in detail by Fisher in the frog, *Xenopus*, and the mouse (1972, 1976, 1979). Over the whole period of retina development in the frog, conventional and ribbon synapses are constantly added, while the number of serial synapses rises suddenly with the onset of metamorphosis. This sudden rise can be prematurely elicited by administration of thyroxine. In *Xenopus*, there is no such increase, but the number of serial synapses remains at a low level throughout metamorphosis. The number of conventional synapses increases — again differing from the results in *Rana* — in two phases. In the first, which ends with stage 47, a rapid accumulation of synapses is found (27 synapses/amacrine cell/h), while in the second phase only slow addition occurs (3.6/am. cell/h). Ribbon synapses have a monophasic increase of 13.7/bipolar cell/h. While they confirm the results for the serial synapses, Tucker and Hollyfield (1977), studying synaptic development under different light conditions, found a slight decrease in conventional and a striking decrease in ribbon synapses, both occurring late in tadpoles reared in a light-dark cycle. In the mouse, Fisher found an initial phase of rapidly increasing synapse numbers, which was followed by a plateau phase. The increase of conventional synapses occurs at a rate of $0.4-1.2/1000 \ \mu m^3/h$, while the rate of ribbon synapse increase is markedly lower. A differentiating amacrine, interplexiform, or bipolar cell forms one new synapse every 2.5 h, in later phases every 45 min. The density of serial synapses remains low.

An influence of light or darkness on the number of synapses in the early stages of the inner plexiform layer development was shown for *Tilapia* (Grün 1979).

In the rabbit (McArdle et al. 1977) the number of conventional synapses at first shows a slow increase followed by a rapid phase and a final plateau. The number of ribbon synapses parallels the rapid phase and the final leveling off. The rapid phase is found between day 9 after birth, when according to Raviola and Raviola (1962) the rabbit receptor cells have developed all elements, and day 18, when the aspect of the retina is adultlike.

In the retina of the adult goldfish, the number of synapses in the central part still increases at a constant rate (Fisher and Easter 1979), thus presumably contributing to enhancement of acuity.

The striking species differences in developmental rates of different types of synapses suggest an underlying functional significance. On different developmental stages different ratios of conventional to ribbon synapses are observed. Though the consequences of these different ratios for visual abilities are not fully understood, some inferences may be given here. A high share of conventional (= amacrine cell) synapses indicates a reduced degree of direct receptor cell-ganglion cell information transfer and more intraretinal processing before the information is conducted to the optic centers. A high share of ribbon synapses indicates a low degree of processing in the ("primitive" or secondarily simplified?) retina and a stressed importance of optic centers in the brain. If this is of any importance to visual characteristics, there should be developmental differences in visual abilities, raised by different ratios of synapse types.

In tissue cultures of the chick embryonic retina, Ruffolo et al. (1978) observed that certain cells are capable of making synapses onto muscle cells over a short period. At the time when this ability no longer exists the already differentiated synapses disappear. Apparently, synapses are produced in an exceeding amount before specification takes place. Those synapses which are not used at the time of specification (e.g., those onto muscle cells in combined tissue cultures) disappear then. The development of the retina inner plexiform layer thus provides an illustration of the "specification by selection," postulated by Changeux and Danchin (1976).

3.5 Neurochemical Differentiation

Neurochemical research in the adult retina is far from being complete, and the increasing amount of literature on this subject, worth a review of its own, is more likely to complicate the picture rather than clarify it. For detailed reviews see Stell (1972), Rodieck (1973), and Graham (1974). A thorough discussion is also provided by Lam (1976).

3.5.1 The Cholinergic System

A way of demonstrating the occurrence of the cholinergic system is the cytochemical localization of acetylcholinesterase (AChE). This method, however, does not always result in a specific proof, but may include other cholinesterases as well. In the developmental stages of the frog (Boell et al. 1955) and the chick (Shen et al. 1956), the ganglion, amacrine, bipolar, and horizontal cells were reported to display AChE activity. This was confirmed by Spira (1974, 1976) for the rat and the guinea pig, where activity was mainly found in the endoplasmic reticulum and in the Golgi system. In the outer plexiform layer of the guinea pig it is scarcely found, but increases with the thickening of the layer. It is not always found at synaptic sites. Lee et al. (1967) localized the first AChE activity in the ganglion cells, and then in the inner plexiform and nuclear layers. The highest activity was always found in ganglion cells. In the inner plexiform layer of the chick, the frog (l. c) as well as in the rabbit (Raviola and Raviola 1962) several bands of AChE were recognized which seemed to be identical with synaptic bands.

A recent parallel to this was the localization of muscarinic acetylcholine receptors in two bands in the differentiating, and three bands in the adult chicken's inner plexiform layer (Sugiyama et al. 1977). For nicotinic ACh receptors, which were observed also in the outer plexiform layer, this was shown to be the case by Vogel and Nirenberg (1976).

Apart from cytochemical studies, nearly all data on developmental changes of the cholinergic system and its constituents have been obtained from the chicken retina.

They show an increase of acetylcholine, which occurs in two steps and is paralleled by an increase of muscarinic ACh receptors and of α-bungarotoxin binding sites.

For details see p. 14. In the *Xenopus* retina, the first appearance of ACh is followed by a 30-fold increase within 48 h (Ma and Grant 1978).

The constituents of the cholinergic system show a gradual increase parallel to the retina development with a peak, not at the "moment" when functional maturity is attained, but at the date of hatching, i.e., when visual function is needed. The decrease in ACh and perhaps the concentration of toxin binding sites subsequently found may then be regulated by the use of the retina cells. A sequence by which the constituents of the cholinergic system appear is not easily established. The synthesis of the ACh receptors is bound to postsynaptic membranes of ganglion, amacrine, and perhaps bipolar cells; the transmitter and the associated enzymes are produced at presynaptic sites. The combined system may be synthesized in the chicken retina for the first time on day 7. If, on the other hand, the presence of one component is indicative of the whole system, one has to assume that as early as day 6 some cells are mature as far as their transmitter system is concerned. In fact, nicotinic ACh receptors have been found to be synthesized before synapses appear (Vogel and Nirenberg 1976). Subsequently, there may be an interdependency of synaptic and cholinergic development.

3.5.2 Aminergic Systems

Catecholamines can be localized cytochemically in synaptic vesicles of the inner plexiform layer in *Tilapia* when the first vesicles appear, i.e., at the very onset of specific retinal differentiation (Grün 1977b). In the chick, the activity of the monoaminoxidase, an enzyme which catalyzes several reactions by which monoaminergic transmitters are degraded, shows an increase between day 12 and day 18, remains at its highest level until 7 days after hatching, and then begins its decrease to an adult level (Suzuki et al. 1977). Thus it shows an interesting parallel to the cholinergic system.

In the rat retina the activity of hydroxyindole-O-methyltransferase (HIOMT), a characteristic enzyme of the indoleamine system, increases sevenfold between day 17 and day 30pp (Cardinali and Rosner 1971); in the chick retina its activity rises 30-fold from day 15 to day 10pp (Wainwright 1979). At the time of hatching a lowering of the increase rate was noted. The role of this enzyme within the retina is unknown. In the retina of the *Xenopus* larva, Baker and Quay (1969) found the 5-hydroxytryptophan-decarboxylase to be present, an enzyme which catalyzes the synthesis of 5-hydroxytryptamine (serotonin). The activity of this enzyme increases at the time of larval hatching, followed by a sudden rise, when the larva begins to swim. At this time, the activity of monoaminoxidase is also very high.

3.5.3 GABA and Other Amino Acids with Transmitter Functions

In the developing retina of all species studied so far, an increase of GABA is noted over a very long period of time. In the mouse (Orr et al. 1976) it lasts from day 2 after birth until day 27, thus differing from glycine (see below). In *Xenopus* (Ma and Grant 1978; Hollyfield et al. 1979) GABA was found to be present prior to acetylcholine and to increase up to metamorphosis, i.e., the increase reaches far into a period after visual function has begun. In the prehatching chick, Pasantes-Morales et al. (1973) found an increase of GABA, paralleled by glutamate and glycine, but contrasting all other free amino acids.

An autoradiographic study by Tunnicliff et al. (1975) shows two transport mechanisms, a low affinity one and a high affinity one, in the chicken retina on day 9. In the 3-week-old chick the high affinity system has disappeared. This suggests a rapid differentiation of GABAergic sites in early phases of retina development, thus differing from what had been found in the rat (Macaione et al. 1970, 1974), where after a steady increase a sharp rise in GABA concentration is reported for later stages. The two transport mechanisms in the chick retina are paralleled by two synthetic pathways, one based on putrescine, and the other starting with glutamic acid; by day 18, only the latter will have remained (De Mello et al. 1976).

Glutamate is not a transmitter of its own but is important as a precursor of other amino acids, particularly GABA.

In the chick (Pasantes-Morales et al. 1973), the mouse (Orr et al. 1976), and the rat (Macaione and Cacioppo 1971; Macaione et al. 1974) glutamate content increases during essential parts of retina differentiation, always paralleling the development of GABA. In the rat, similar behavior was observed for glutamine and glutamosynthetase (Macaione and Cacioppo 1971). According to Riepe and Norenberg (1978; rat) and Norenberg et al. (1980; chick), however, this enzyme occurs exclusively in Müller cells. The glutamate increase in the mouse was found to coincide with the phase of receptor cell synapse development; since, however, no loss of glutamate occurred in dystrophic animals, any causal correlation remains open to question.

Glycine shows an increase in the developing retina of the chick (Pasantes-Morales et al. 1973), the rat (Macaione et al. 1974), and the mouse (Orr et al. 1976), but taurine only in the two rodents. The developmental profiles of thes two amino acids are essentially similar to those of GABA and glutamate. But in the mouse, after an initial rise from day 22pp to day 10 pp, glycine clearly decreases, while taurine begins to increase on day 6pp only.

3.5.4 Gangliosides

The development of gangliosides and phospholipids was studied in the chick retina by Dreyfus et al. (1975). They found three phases of N-acetylneuramine acid (sialic acid) formation, one at the onset of retinal cell differentiation, a second during receptor cell and synapse maturation, and a third beginning only on day 18 posthatch. While the pattern of cellular phospholipids undergoes no changes, stage specific differences in gangliosides are detected.

3.5.5 Cyclic Nucleotide Systems

Even if the cyclic nucleotides should turn out as having a metabolic rather than a specific retinic role in receptor cells, there can be no doubt that a specific function must be assumed for the neural part of the retina, as well as for other parts of the nervous tissue, which justifies a consideration of this metabolic system in this place.

The developmental profile of the cAMP level in the chick retina shows a decrease after day 10, followed by a rise after day 16. The content in cGMP is 30-fold less (Fletcher and Chader 1978; De Mello 1978). In the rat, Farber and Lolley (1977) observed that at the time of synapse formation and functional maturation the cAMP metabolism has come to a state of stability.

The activity of mouse retina adenylate cyclase steadily increases up to day 10pp, when adult level is attained (Lolley et al. 1974). The activity could be stimulated by NaF and dopamine. In the rat, Farber and Lolley (1977) found a fivefold increase of adenylate cyclase up to day 25pp. The main activity of adenylate cyclase is localized in neuronal layers as could be shown in rats where receptor cells had degenerated.

The activity of the ATP-kinase rises in the chick retina between day 18 and hatching, while GTP-kinase starts its increase on day 13 only (Fletcher and Chader 1978). Both enzymes thus parallel the retina differentiation.

3.5.6 Adenosine Triphosphatases

Adenosine triphosphatases in nervous tissues have a transport function (Bonting et al. 1962) as well as a role in neurochemical transmission (Vizi 1978). Apart from this, there is ample evidence for the presence of Na^+-, K^+-, Mg^{2+}-, and Ca^{2+}-dependent ATPases in photoreceptor cell outer segments (Berman et al. 1977; Sack and Harris 1977) where an involvement in specific outer segment activity is assumed (Robinson et al. 1975).

In the chick, three peaks of ATPase were found during retina differentiation (Coulombre 1955): the first when the inner plexiform layer appears, the second when the outer plexiform layer is formed, and the third at the time of receptor outer segment formation. A further increase in ATPase activity was observed by Yew et al. (1975) at the time of hatching, i.e., at the time of beginning function.

In dystrophic C3H mice, the amount of ATPase decreases, parallel to the degeneration of receptor cells after day 20pp (Lolley and Racz 1972).

An ultracytochemical study in *Tilapia* (Grün 1977a) shows ATPase to appear in the inner plexiform layer, synchronously with immature synapses. The enzyme is also found in synaptic vesicles as soon as these appear, and in inner and outer segments with the onset of their specific differentiation. In all cases mentioned here, it remains to be clarified whether a peak or an increase in ATPase causes, is caused by, or is merely synchronous with the maturation of the specific functional sites.

3.6 Development of Retinomotor Response

Retinomotor response is the characteristic stretching reaction to light, which makes the rods attain a position of their outer segments where these are separated from neighboring outer segments by processes of the pigmented cells. With decreasing light intensity or the beginning of darkness, the inner and outer segments are withdrawn.

Occasionally under this heading, the migration of pigment granules within the pigment cell processes is also covered. In the chick (Hasama 1941), pigment migration begins shortly after hatching and increases in intensity up to 2 months of age. It could be elicited by sunlight, ultraviolet, and infrared light, but not by X- or radium (γ) rays. In the cichlid *Tilapia leucosticta* pigment migration was found 1 day after all receptor types are present and seem to be functional (G. Grün, unpublished work).

All studies concerned with the development of receptor cell stretching or contraction have been performed in fish species. Retinomotor responses appear after receptor cells (single cones and rods) have differentiated. In *Gobius* (Eramisheva 1956) there is a delay of 1 month between the cone development and the appearance of retinomotor reaction. In the pacific salmon, *Oncorhynchus*, Ali (1959) found cones already in the embryo, but it was only in the alevin that myoid contraction was noted, and only in the fry that an immediate and distinct light reaction was apparent. In the cichlid *Nannacara anomala*, Wagner (1974) noted the first retinomotor responses on day 16, while on day 5 cones are differentiated and the stage of free swimming has been attained. Rods, however, are not found before day 16. Finally, Blaxter and Jones (1967) and Blaxter and Staines (1970) noted a temporal correlation of the appearance of retinomotor reaction and outer segment development in rods in the herring and ten further species. This is mostly simultaneous to metamorphosis. It is assumed that also in the species mentioned above where retinomotor appearance has been correlated to cone development, its late appearance may be caused by the late development of rods.

Armstrong (1964), in his study on the bullhead (*Ictalurus nebulosus*) retina development, came to the conclusion that the mechanism of retinomotor response is already present before, and awaits, receptor cell function. It is then triggered by this function. In fact, tubuli formation in receptor myoids is often noted rather early in development, thus giving an example of "function at rest," awaiting its releasing influence.

3.7 Functional Development of the Ganglion Cells

Crucial to the understanding of the ganglion cell functioning as a unit is the concept of receptive fields. Morphologically, this area seems to be the extent of all dendritic endings of this single cell. Physiologically it is characterized not only by its extent and size, but also by its organization and by the type of stimulating light spot upon which a reaction is found. For further details see Levick (1972) and Rodieck (1973). The development of these physiological properties and their correlation to dendrite morphology (see Levick 1975) will be discussed in the following.

In a certain type of ganglion cells of the frog *Rana temporaria* – class 2 cells, which are thought to be sensitive to moving stimuli – Reuter (1969) found a change of dendritic morphology during metamorphosis, i.e., rather late in comparison to other

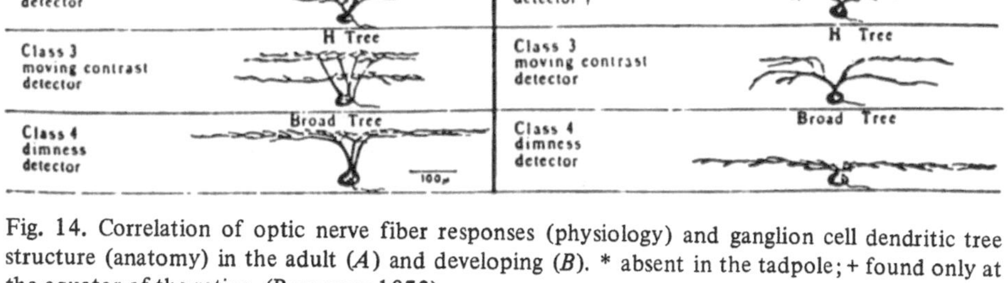

Fig. 14. Correlation of optic nerve fiber responses (physiology) and ganglion cell dendritic tree structure (anatomy) in the adult (*A*) and developing (*B*). * absent in the tadpole; + found only at the equator of the retina. (Pomeranz 1972)

developmental events. According to Pomeranz and Chung (1970) and Pomeranz (1972; see Fig. 14), in the tadpole retina class 1 ganglion cells (morphological: constricted tree; physiologically: edge detector) are lacking; class 2 ganglion cells (E-tree; convex-edge detector) are not found in peripheral zones; only class 3 (H-tree; moving contrast) and class 4 ganglion cells (broad tree; dimness detector) are present all over the retina. In the adult frog retina, all four types can be localized. Chung et al. (1975) revealed similar relationships in the *Xenopus* larva. After stage 43 the majority of ganglion cells show "on" responses; in other cells, "off" responses can be elicited. After stage 47 an additional response to dimming light is found. Up to metamorphosis, 35% of the ganglion cells are found to be event units, reacting either to "light on" or to "light off," 37% are dimming units, and other ganglion cells are of a mixed type. To the development of the two main functional types, two developing morphologically types may be paralleled, T-cells and H-cells. Shortly before metamorphosis, bushy ganglion cells are formed, synchronous with the appearance of sustained responses.

The same authors report a stepwise decrease of receptive field size, while concomitantly the branching pattern of the dendrites becomes more and more complicated. Rusoff and Dubin (1977) observed a differential development of receptive field size in the retina of the cat. Between 3 and 4 weeks pp the size of the centers is reduced to the adult value of $3°$, corresponding to a dendritic field size of 300 μm. The size of the surrounds, however, is larger than that of the adults at least as far as 7 weeks pp. While thus the whole receptive field is larger than in the adult, the dendritic extent at 3 weeks is already of adult size (Rusoff and Dubin 1978). It is assumed that larger receptive fields are formed by specific synaptic connections which disappear later. This seems to be of importance for eyes which grow all through life, as is the case in teleosts. In the adult goldfish (Easter et al. 1977) the size of receptive fields is unchanged, regardless of the animal's and the eye's changing sizes.

In the 9-day-old rabbit, Bowe-Anders et al. (1975) found no ganglion cell receptive fields with a center surround organization. By day 21pp antagonistic surrounds were localized in all on-off cells. From day 11pp on, they found cells which were sensitive to moving stimuli. In the same species, Masland (1977) localized concentrically organized and directional selective receptive fields on day 10pp, which, however, revealed clear differences to adult retinae. The cells were of the on-, off-, or on-off type. Adult organization developed between day 10pp and day 20pp.

All the events looked at so far have in common that they occur relatively late when compared to other topics of retina development. All structures necessary for retina function have been formed before and the retinae must be assumed to be functionally mature. Stages of a fine adjustment are postponed to the development of specific activity. Thus the retina appears to be functioning imperfectly or at a lower level of organization before this final adjustment takes place. This does not necessarily express a state of lesser adaptation in larvae or other early stages, but probably reflects an adaptedness to lesser needs. When only after the attainment of a certain developmental age a certain receptive field organization or the sensitivity to moving or other specific stimuli appears, this means that up to this stage the animal could well do without these properties. In the framework of their simpler behavioral patterns they had no need for these properties, adapted, as they were, to life under the mother's care or the largely constricted conditions of a tadpole.

A correlation of myelination of the optic tract and developmental events of physiological importance have scarcely been reported. Tilney and Casamajor (1924) found myelination to occur synchronously with eye-opening in 6-day-old kittens. In *Xenopus* both the beginning of myelination and the onset of visually guided behavior occur at stage 49 (Ganze and Peters 1961; Wilson 1971). In the chick, myelination begins with day 15 (Rager 1976), thus paralleling the maturation of receptor cells and the retina's primitive function. A direct interdependency between myelination and other events cannot be made out at present, and it may well result that the correlations shown here reflect an unknown common cause, or are mere chance.

Evoked potentials — as aroused by electrical, not visual stimulation of the tractus opticus — have been found in the optic centers of the chick (Blozovsky 1971a, b; Rager 1976) and the rabbit (Hunt and Goldring 1951) before light evoked potentials could be recorded (Raviola and Raviola 1962, in the rabbit) and even before retinal functional sites have matured. This observation leaves no doubt that ganglion cells, or at least their axons, are ready to function before being active, and that it is not the optic tract where the final steps in retina maturation have to be looked for.

3.8 Development of the Electroretinogram

The electroretinogram (ERG) displays several peaks or waves. These are mainly the fast cornea-negative a-wave (= PIII) and the fast cornea-positive b-wave (PII). In addition, a slow cornea-positive c-wave (PI) is often found, and in some cases, a positive d-wave. For further details see Rodieck (1973) and Tomita (1978).

An ontogenetic aspect of the ERG is found in the appearance of its components, its size, and its threshold.

The question which of the components or waves appears first must be answered differently for different species. A cornea-negative deflection, mostly designated as a-wave, as a first sign of an ERG has been found in the frog, *Xenopus*, the dog, the rabbit, and the rat, and is probable for the chick.

For the rat, slightly differing results can be presented, see p. 24. In the rabbit, its appearance has been recorded at day 6pp to day 8pp. A positive potential is found by d10pp, or earlier when an intense illumination is applied. It seems to consist of two components which later form the adult b-wave (Noell 1958; Masland 1977). The observation of Bonaventure et al. (1967) that at day 10pp a biphasic ERG appears fits well to these data. With proceeding development, the relative share of the negative potential decreases. For day 18pp the existence of an a-, b-, and a c-wave is reported (Demirchoglian and Mirzoian 1953).

In the dog, Horsten and Winkelmann (1960) report a negative wave for d10pp, followed by a rapidly increasing b-wave on d15pp.

For three species of the frog (*Rana*) the sequence has been found to be the same: in *R. temporaria* (Müller-Limmroth and Andrée 1954) a negative potential is found at first, and 1 week later a-, b-, and c-waves are recorded. In *R. pipiens* (Nilsson and Crescitelli 1970) and in *R. catesbeiana* (Nilsson and Crescitelli 1969), the cornea-negative potential is very slow, but some days later, rapid waves appear. Finally positive waves are found, which more and more become the dominating ones. For *R. pipiens*, this result was confirmed by Lam (1977). In *Xenopus laevis*, a negative wave (PIII) can consistently be demonstrated from stage 39 on (Witkovsky et al. 1976). From stage 40 on, i.e., 10 h later, the b-wave is seen; it increases rapidly and is soon the predominant one. A positive deflection thought to be a c-wave is found at intense illumination from st. 58 on.

In the chick, more studies have been done than in any other species. Summarizing the data given on p. 16, we find a purely negative ERG, on days 16—18, and an onset with a biphasic ERG on days 17, 18, or 19. As far as the data for first recording are concerned, differences may be due to heterogeneities of rearing methods, races, or individuals, and the chick may well be classified among those animals beginning with a negative wave.

An electroretinogram beginning with a positive potential has been reported for one mammal and one urodelan species only. In the 6- to 10-day-old cat, Zetterström (1956) found an ERG which consisted of a positive deflection only. No a-wave was recorded. In one case, however — a kitten which was too deeply anesthetized and subsequently died — only a negative response was found. It can be assumed that this negative wave is normally present but is covered by the positive one. In accordance with this finding, Tucker et al. (1979) observed the late receptor potential (= the ERG component giving rise to the recorded a-wave) to appear on day 9.

In the premetamorphic stages of *Salamandra salamandra*, no a- and no d-waves were recorded (Himstedt 1970). First appearance of a cornea-positive wave or a biphasic ERG was also noted in the regenerating eye of an adult newt (Lam 1977). The third type of ERG differentiation, where, at first recording, the electroretinogram is found to be biphasic and composed of a positive and a negative deflection, has been found in rodents and primates. (See, however, the chick, p. 16). In the very first study on ERG development, Keeler et al. (1928) showed that by most intense illumination a negative reaction was found in the 13-day-old mouse, while a positive one was only indicated. Both waves had a low and stretched shape. Noell (1958) made the same observation for day 12pp. On day 14pp, a distinct b-wave is accompanied by a clear a-wave; but as late as day 18pp the latter could not easily be found at low stimulus intensities.

Similarly, in the guinea pig Bornschein (1959) found a positive deflection at birth; a negative one was found only when intense illumination was applied.

In the rhesus monkey, *Macaca mulatta*, an "a+b-complex" is found on the 2nd day after birth (Ordy et al. 1962). Also the human retina must be mentioned in this group. Horsten and Winkelman (1962) could separate an a- and a b-wave in the newborn infant. This biphasic ERG was already recorded in the 7.5-month-old fetus by Samson-Dollfus (1968).

There is a reason for the suspicion that the electroretinograms recorded at birth do not represent the very first ones. In the guinea pig, the ERG at birth is said to be a rather advanced one, and fetal stages, which have not been studied, may reveal an immature ERG. The same is assumed for the rhesus monkey. In the mouse, however, the retina does not appear to be so differentiated before day 12pp as to allow an ERG to be produced. At present, we are left with the assumption that from the three modes of ERG development displayed here two may have to be abandoned in favor of the one according to which the ERG starts its appearance with a negative deflection. But the results may vary with the kinds of stimuli, light intensities and techniques applied, and one cannot draw conclusions from one or two observations.

Besides composition and shape, there are other properties of the electroretinogram which are subject to developmental change. A most general observation is that of an increase in "size", or amplitude. It has been reported for all species mentioned so far as well as for the duck (Paulson 1965). Since this increase occurs gradually and parallel to retina development (e.g., in the rat: Dowling and Sidman 1962; see p. 24), it may be assumed that it is due to the peripheral proceeding of retina maturation which involves increasing numbers of cells.

A long latency is characteristic of the ERG of the mouse (Noell 1958) and the chick (Garcia-Austt and Patetta-Queirolo 1961), where it is reduced from 40 ms on day 18 to 8 ms on day 20, and to 5.5 ms on day 6pp (Blozovsky and Blozovsky 1968). For the rabbit, however, a shorter latency is reported for the earlier stages, which increases later (Bonaventure et al. 1967). Furthermore, an increase in sensitivity or a lowering of the threshold is generally noted. This also may be seen in connection with the maturation of increasing numbers of cells.

The appearance of the ERG components gives an indication of the developmental events in the retina, based on the topographical origin of the components (see Tomita 1978; Rodieck 1973). Horsten and Winkelman (1962) found no correlation between the appearance or nonappearance of an electroretinogram and any histological events in the dog. Clinical cases of human retinae were found by the same authors to be structurally normal. In other cases (rabbit: Noell 1958; Raviola and Raviola 1962; rat: Bonting et al. 1961) as well as by application of the electron microscope (rat: Dowling and Sidman 1962; *Xenopus*: Witkovsky et al. 1976) correlations are distinctly visible between the appearance of an ERG, the maturation of receptor cell inner segments, and the formation of receptor cell outer segments, as well as between an increase in ERG size and growth of outer segment length.

A more detailed study has been done in *Rana catesbeiana* (Nilsson and Crescitelli 1969) and *R. pipiens* (Nilsson and Crescitelli 1970). In both species, the appearance of a first negative wave was found to be synchronous with the formation of outer segment disks. This confirmed an observation in the rabbit (Noell 1958). A second, fast negative response was recorded when a receptor terminal appeared, with still immature synapses, however. When a considerable number of mature synapses was present in the terminals, a first positive wave was noted. Similar correlations were found for the development of the *Xenopus* ERG by Witkovsky et al. (1976).

The negative waves are thus correlated to receptor cell development, while the positive wave is not primarily bound to the mature terminal synapse, but to the formation of contacts with cells of the inner neuronal layer (Müller cells) where the b-wave is said to take its origin. Correspondingly, Blanks et al. (1974b) found the b-wave even after degeneration of the receptor cells and their terminal in rd mice. Rager (1979) provided evidence that the appearance of the b-wave is correlated to structural maturation of the glial Müller cells.

There have been few attempts to obtain information on the influence of extrinsic factors on ERG development. According to Bornschein (1959), the presence of light is not necessary, because the guinea pig at birth has a rather far developed ERG; it must have been developing in darkness. The same can be assumed for the rhesus monkey and man. But, even if not necessary, the presence or absence of light may have a modifying effect on ERG formation. In the cat, where essential parts of retina development take place after birth, a delay is observed when the kittens are born and bred in absolute darkness (Zetterström 1956). Reuter (1976) could not confirm this for the rabbit. Considerable delay of ERG appearance and maturation in *Xenopus* is caused by artificially induced vitamin A deficiency of the mother and the tadpoles (Witkovsky et al. 1976).

4 Development of the Retina by Differentiation of Single Cells

When it is possible that adult function of the whole retina is mainly achieved by the realization of a program which is inherent to single cells or cell types, without a major form of extracellular regulation or coordination, then this should be expected to occur in isolated cells too. Two ways of isolating cells shall be considered in this respect, the development in vitro, and a "theoretical isolation" of cells.

4.1 In Vitro Development of the Retina

The purpose of this section is to obtain an impression of the extent to which the retina cells are capable of differentiating when separated from their normal environment. Since only this aspect of in vitro differentiation is considered, this survey is not complete. Above all, it must be mentioned that the studies by Moscona and colleagues, which are based on quite different intentions, are omitted here. A recent review on this aspect has been presented by Moscona (1976). The initial intention was to show the inherent capabilities of self-differentiation in single cells. However, no single cell culturing has been done in the retina so far, and one has to rely on tissue culture studies or cultures of dissociated cells. These of course have a cellular surrounding and a quasi normal environment, and it has repeatedly been shown (e.g., Monard et al. 1973) that structural differentiation may be induced by glial cells. But, as Hild and Callas (1967) noted, "the explanted retina does not behave as an organized entity . . . developmental differentiation was observed only at the cellular level."

The development of retinae in vitro has been observed mainly in the chick. Some results obtained in this species and in the rat will be combined to give a picture of developmental potentials within the retina cells.

Explanted retina cells, after having been separated from the pigmented epithelium, first form aggregates and rosettes (Stefanelli et al. 1961c; Hild and Callas 1967; Sheffield 1970; Sheffield and Moscona 1970; Crisanti-Combes et al. 1977), which had already been observed in the rat by Tansley (1933). This formation of rosettes may be mediated by a cell-adhesion molecule which is involved in the normal development of histological layers in the chick retina (Buskirk et al. 1980). In monolayer cultures, Müller cells form a substratum upon which neuroblasts differentiate further (Cristanti-Combes et al. 1977). Apparently all types of retina cells can differentiate under culture conditions (Hild and Callas 1967; LaVail and Hild 1971; Fujisawa et al. 1974). Ganglion cell neurites grow out (Meyer 1936; Crisanti-Combes et al. 1977; Ulshafer and Clavert 1980), and it was by in vitro observation that the factors for guidance and orientation of the optic tract were found to be confined to the ganglion cell area (Goldberg 1977). The outgrowing fibers of the goldfish retina are said to have a tendency toward clockwise directionality, which suggests an inherent helicity (Heacock and Agranoff 1977). In explanted retina pieces their growth and direction, at least in initial phases, is independent of external stimuli (Johns et al. 1978). McLoon and Hughes (1978), however, found an increasing degeneration of explanted optic fibers and ganglion cells at the time when normally central connections are formed. Presumably the absence of the target led to the degenerative development.

Neuropil with axonal and dendritic processes is regularly found and usually regarded as forming part of the inner plexiform layer (Stefanelli et al. 1961b, 1967b; Hild and Callas 1967; Fujisawa et al. 1974). It is formed within rosettes (Stefanelli et al. 1961b; Sheffield and Moscona 1970). Besides several types of junctions (Sheffield 1970; Sheffield and Moscona 1970) mature synaptic contacts are found (Stefanelli et al. 1966b; Fig. 15a), some of them with synaptic ribbons (Crisanti-Combes et al. 1977; LaVail and Hild 1971). Vogel et al. (1976) recognized at least three types of synapses, similar to those of the in-ovo-retina in an amount of 10^8/mg protein. Furthermore, nicotinic acetylcholine receptors were found to differentiate in the dissociated embryonic retina. Similar to what has been found in vivo, these receptors develop independently of neuron differentiation (Vogel and Nirenberg 1976; Vogel et al. 1976).

Another component of the cholinergic system, the choline acetyltransferase activity, was found to behave in vitro similar to that in ovo (Crisanti-Combes et al. 1978). Ramirez (1977), who found far-reaching structural differentiation of the retina in vitro, and a high degree of complexity, stated that from a cholinergic point of view the retina is rather self-sufficient (in contrast to later stages of the tectum opticum). Horizontal cells and complete receptor terminals, as characterized by the occurrence of synaptic ribbons, have been found (Stefanelli et al. 1966a, 1967a; LaVail and Hild 1971; Crisanti-Combes et al. 1978). This was not the case in cultures grown on the chorioallantois (Fujisawa et al. 1974).

The formation of green, yellow, and red oil droplets in cultured retina (Yata 1961, cited after Cooper and Meyer 1968; Govardovsky and Kharkeevich 1965) not only marks the appearance of receptor cell inner segments but also shows that the cells differentiate into several types of cones, i.e., accessory and main element of double cones and single cones. In the rat and the mouse, further inner segment characteristics are numerous free ribosomes, mitochondria, ER, a Golgi apparatus, terminal bars, centrioles, and a connecting cilium (Hild and Callas 1967; Tamai et al. 1978), and in the chick, a cluster of vesicles and lamellae, forming what might be called a paraboloid (Stefanelli et al. 1966a).

Receptor cell outer segments have not been observed to develop in vitro (Fujisawa et al. 1974; Crisanti-Combes et al. 1977), or were only indicated by the formation of a lamellar system or a few disks (Stefanelli et al. 1966b, 1967b; Hild and Callas 1967; Barr-Nea and Barishak 1970; LaVail and Hild 1971; Fig. 15b). This is not surprising, since it is known that pigment epithelium cells are necessary for the orderly development of outer segments (Silverstein et al. 1971; Hollyfield and Witkovsky 1974), because the pigmented cells provide for vitamin A, the lack of which leads to incomplete outer segment formation or to degenerated outer segments (Eakin 1964). Sidman (1961b and cited after Lucas 1965) found a stimulation of rod differentiation in vitro when 11-cis-retinol was added, but not when all-trans-retinol was administered; this clearly proves the importance of the pigmented epithelium, where the biochemical transformation takes place. Tamai et al. (1978) confirmed this by demonstrating that in combined cultures of previous separated retina and pigmented epithelia a moderate amount of lamellar membranes was produced which were arrayed parallel to the long axis of the cilium.

The retina appears as a tissue with a far-reaching potential of self-differentiation. Apart from outer segments and, to a certain degree, optic fibers, all structures which are important for specific functions develop to a very high degree and apparently attain maturity. A high intrinsic ability of differentiation along with an "organizing po-

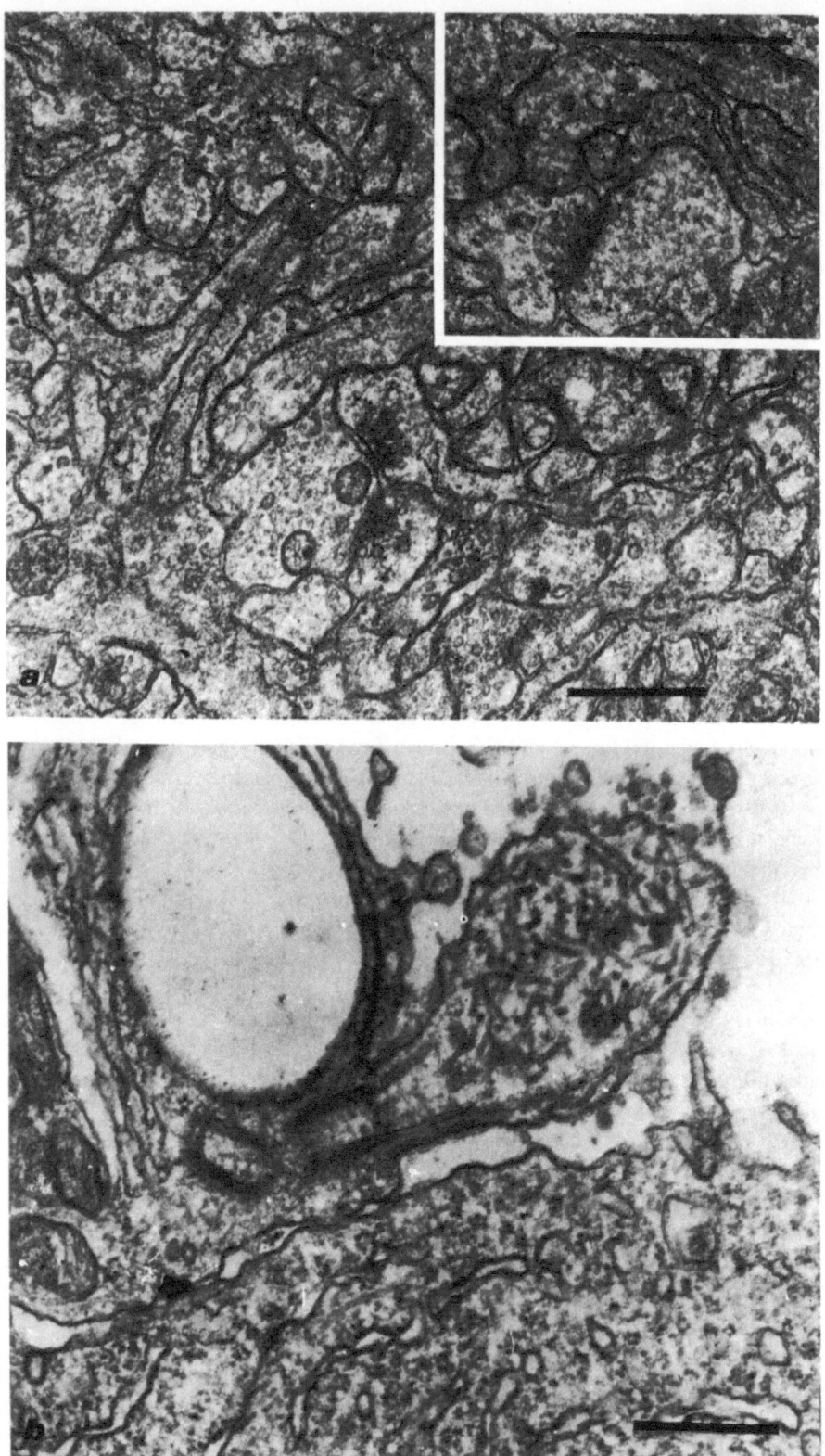

tential" (Ramirez 1977) can be assumed, which allows an intercellular differentiation and the formation of a complicated network. This result is taken as an indication that tissue organization is achieved by way of intracellular development.

4.2 Intracellular Differentiation

A way of isolating cells theoretically will be tried in the following. From the data of the six species, all those are combined which can be regarded as occurring within single cells. This is to say that such processes as layering, an overall increase of any neurochemical component, the appearance of electrical activity, etc. – processes which only can be conceived and observed over a greater fraction of the retina – are not included. This is not to say that the causal relationships also are strictly intracellular, but the intention is that the resulting differentiational steps are performed by single cells, forming nonetheless – this is the concept – a functioning retina.

The data selected for this purpose will be arranged according to cell types (ganglion cells, amacrine cells, etc.) and are pooled for all species in question. This requires of course a common time scale. In Table 1 a time scale is proposed where the days and weeks of the various developmental periods are converted into developmental units, DU. These were determined by considering the beginning, the end, and several landmarks of retinal development, as well as the number of hours, days, or weeks which arbitrarily subdivide the flow of development.

Table 1. Conversion of the various calendarian units into developmental units (DU). The obvious incongruencies found in this table reflect the existing differences concerning the sequence and the temporal distances of various events. Therefore this table is only a first approach. P – Partus. [a] larval hatch. [b] hatch from mother's mouth

DU	Tilapia d	Xenopus d	st.	Gallus d	Mus d	Rattus d	Homo w
1	4	2	28–34	2– 5	11–17	14–16	4– 8
2	5[a]	3	35–37	5–10	17–P	16–P	6–11
3	6	3[a]	37–38	8–11	P–5pp	P–3pp	8–11
4	7	3	39–40	10–16	1– 7pp	3– 8pp	12
5	8	4	41–43	16	7–10pp	8–12pp	12
6	9	4/5	44–46	16–	10–13pp	12–15pp	to
7	10–11	5/6	46–47	hatch	13–		32
8	11–12		47 to				
9	13[b]		meta-	0– 2pp	16–24pp	15–21pp	32–
10	13		morpho-	2pp	to 30pp	21pp	–P
			sis				

Fig. 15. a Chick embryo retina after 25 days in vitro. Reaggregated cells have produced a plexiform layer containing synapses. Bars = 0.5 μm. (Stefanelli et al. 1966a). b Chick embryo retina after 28 days in culture. Centrioles and ciliary stalk of an outer segment. The tubules in the stalk show a great degree of disorder. Bar = 0.5 μm. (Stefanelli et al. 1967a)

61

4.3 Ganglion Cell

In DU 1 the last mitosis is accomplished and the ganglion cell migrates to the vitreal border, becoming spheroidal. By the settling at a definite site, its topographical relationship with the tectum opticum has become specified. The cell now increases in volume, producing more mitochondria, rER, RNA, and acid phosphatase. The cell becomes unipolar, indicated by the increasing activity of diaphorase in its vitreal part and the outgrowth of a neurite.

In DU 2, this outgrowth has developed neurotubuli, neurofilaments, and mitochondria. Axonal transport is directed from the soma to the terminal. At the end of this DU, the neurite has reached the optic center. At the scleral part of the ganglion cell a conical protrusion announces the formation of a dendrite. A Golgi system appears within the cytoplasm and ER profiles become more numerous. Carboxylesterase is produced in detectable amounts. Somewhat later, in DU 3, when dendrites have grown and begin to produce horizontally oriented arborizations, diaphorase is found in the scleral part too, and acetylcholinesterase can be demonstrated in the cytoplasm or on the cell membrane. The neurite, which suddenly acquires the ability to bind the lectin concanavalin A, is now also capable of evoking a reaction in the optic center upon electrical stimulation, but is not yet myelinated. Obviously it can conduct impulses and has formed synapses.

Subsurface cisterns are produced in DU 4. A certain state of stability has been achieved, because at the end of DU 4, the cell can no longer be destroyed by an irradiation with 500 R. Furthermore, in DU 4, Golgi impregnation is successfully applied.

In DU 5, a number of oxidative enzymes are demonstrably present, and the cell passes another developmental step, after which, in DU 6, it can be classified morphologically and physiologically. It is only after this date, in DU 7, that the myelination of the neurite begins, proceeding from the optic center to the perikaryon. Simultaneously, mucopolysaccharides appear, and Nissl substances and alkaline phosphatase are found later. The latter might be taken as an indication of a terminated development. In DU 8, the completion of differentiation is shown by the final pattern of certain enzymes. After DU 8, the size of the receptive field decreases.

4.4 Inner Plexiform Layer

In view of the difficulties in attributing components of the synaptic layers to a certain type of cells, the neuropil-forming parts of ganglion, amacrine, and bipolar cells are treated here as if they were a separate cell. All data in this section represent intracellular events.

In DU 2, tangentially oriented processes appear, and one DU later, their membranes bind concanavalin A, and specialized membrane sites are formed, near to which synaptic vesicles may be found. Neurotubules are a common feature now. In DU 4, the membrane sites have differentiated into true synaptic junctions. In the core of the few synaptic vesicles and along the membranes of the axon terminals, adenosine triphosphatase can be localized now. Beginning with DU 5, the number of synapses increases, mostly by passing discrete phases. The amount and the localization of ATPase is now adultlike. Part of the synaptic vesicles show positive reaction to cytochemical staining of catecholamines. Both acid and alkaline phosphatase are present. In DU 7, the ratio of mature synapses to immature synapses increases exponentially.

4.5 Amacrine Cell

Similar to the ganglion cell, the amacrine cell accomplishes its last cell cycle in DU 1. It passes a bipolar phase and migrates into the ganglion cell layer. Leaving this layer, it looses the axon and settles in its final region. Only then are definite processes formed. Many ribosomes are in the cytoplasm, and ER and mitochondria mark the onset of differentiational events. Golgi systems join them in DU 2. In the course of this DU, and consequently synchronous with ganglion cell dendrite formation, the amacrine cell sends out one or more axonal processes. But only in DU 3 and 4 do these branch horizontally and form an axon terminal, synaptic vesicles, and synaptic junctions.

Subsurface cisterns are found in the perikaryon by DU 4, when the amacrine cell is no longer harmed by 500 R, and consequently is assumed to have achieved a stabilized developmental state.

Up to DU 5, the amacrine cell has to establish final contact to a bipolar cell or another amacrine cell. The localization of ATPase within the terminals is now adult-like. It is only in DU 8 that the amacrine cell produces Nissl substance.

4.6 Bipolar Cell

Unlike the hitherto mentioned cells, the bipolar cell completes its last cell cycle during DU 2 only. During this and the following DU, there are very many ribosomes in the cytoplasm; mitochondria, ER profiles, and finally single Golgi zones appear. A neurite grows out of a vitreal axon hillock, where ribosomes and neurofilaments are localized. DU 4 is characterized by a decrease of ribosomes and mitochondria. The axon has penetrated deep into the inner plexiform layer and forms a terminal and typical ribbon synapses. Synchronously, a dendrite grows out of the scleral part of the cell, invaginates a receptor terminal and presumably forms a synapse.

From this DU on, the cell can be recognized in Golgi impregnations. The above-mentioned decrease in ribosome number leads to a lowering in RNA concentration which is finally lower than that of the ganglion and the amacrine cell.

Irradiation with 500 R renders the bipolar cell pycnotic. This effect is lost in DU 5, when acid phosphatase and diaphorase can be demonstrated. There is no further change up to DU 7, when mucopolysaccharides appear, and DU 8, when Nissl bodies and alkaline phosphatase indicate an end of differentiation, as does the final distribution of certain enzymes.

4.7 Horizontal Cell

Unlike the bipolar cell, but similar to the amacrine and ganglion cell, the horizontal cell completes its last cell cycle in DU 1. Up to DU 3, it has produced processes and made contacts to bipolar cell processes. Diaphorase and acetylcholinesterase appear; the latter, however, disappears up to DU 5. In DU 4 or 5, but usually prior to the bipolar cells, the horizontal cell dendrites invaginate a receptor cell terminal. Only in DU 6 are the junctions between the horizontal cell and the bipolar cell mature. In DU 7, the horizontal cell can be called mature.

4.8 Receptor Cell

The receptor cell, too, finishes its last cell cycle in DU 1. It contains many ribosomes and mitochondria in increasing numbers. In the apical part of the receptor cell, two centrioles face each other at right angles, and here microvilli are formed as well as junctions with the pigmented epithelium. Gap junctions with neighboring cells are produced in DU 2.

Now the cell contains large amounts of RNA. Golgi systems are found, and scleral to the nucleus, mitochondria begin to accumulate. The determination of the receptor cell becomes visible: A chief element of a double cone now begins to synthesize the material for its oil droplet. If it becomes the accessory element, another substance, galloxanthine, will be stored beginning with the following DU. In both cases, however, no oil droplet will be visible yet. In DU 3, the inner segment surpasses the outer limiting membrane and begins to form a ciliary outgrowth. At the other pole of the cell, a receptor terminal is formed by stretching of the distal cell part.

In DU 4, a horizontal cell process superficially penetrates into this terminal, and some synaptic vesicles as well as precursors of synaptic lamellae appear. If the cell is a member of a double cone, subsurface cisterns are produced in the inner segment by fusion of flattened vesicles. The inner segment has now been filled with mitochondria, and Golgi vesicles and ER profiles are more numerous. The scleral part of the ciliary outgrowth is strangely blown up. Toward the end of DU 4, the synaptic ribbons have elongated up to 1000 nm or more, and a synaptic spindle is formed. Although the differentiation seems rather advanced, irradiation of the cell by means of 500-R X-rays leads to pycnosis and degeneration. It is only at the end of this DU that the receptor cell is not subject to any effect of 500 R.

This is also the time when the formation of outer segments begins. The number of disks increases, and when a certain number has been produced in DU 5, it is possible to detect visual pigment. Now, almost synchronously, in a rod receptor cell, phagosomes begin to be shed and triadic configuration in the terminals as well as mature synapses are present. All this suggests a beginning of the cell's function.

The synaptic vesicles have achieved their final distribution and number. ATPase is present in outer segment disks, diaphorase in the inner segments, as well as weak PAS and mucopolysaccharide specific staining. Now the oil droplet can be seen vitreal to the mitochondria. A paraboloid is developing in the inner segment.

DU 6 and 7 are characterized by enhanced growth of the outer segment. If the cell is a single cone, small astaxanthine droplets will now fuse to form a single oil droplet. In the apical part of the cell, junctions with Müller cells are formed. At the end of this DU, the cone paraboloid has completed its development and stores glycogen. The staining of PAS-positive substances shows no further changes. If the cell is a rod, the latter process will occur in DU 8, when the outer segment attains adult length.

This compilation of cell autonomous processes comprises essential features and can serve as part of a design of a generalized scheme of vertebrate retina development. It becomes evident that, by realizing its inherent programme, each cell achieves a state of adult activities. This does not mean that the cell develops independently of its environment.

On the contrary, it has long been known that, e.g., the presence of pigmented epithelium is required for a receptor cell to differentiate (see above) and even seems to provide positional information to ganglion cells (Levine 1979). Similar conditions may exist for other cells. The intention is to show that the organized tissue "retina" matures by mere realization of cellular programmes without superimposed coordination.

This state of primitive maturity, however, does not represent true retina function. Final maturity is only brought about by additional developmental processes which are not observed at cellular, but only at tissue level. This other part of a generalized scheme of retina development will be presented in the following.

5 Differentiation at Tissue Level

5.1 Proliferation

Amacrine and ganglion cells finish their last cell cycle in DU 1, but at the scleral border of the retina, there are still mitoses to be found. This does not change up to DU 4, when the number of mitoses decreases. In DU 5, there are no further mitoses.

5.2 Layering and General Differentiation

Prior to DU 1 the optic cup and the optic fissure are formed; the future retina consists of an undifferentiated epithelium, where cells are interconnected by zonulae adhaerentes, macula adhaerentia, and gap junctions. At the vitreal and the scleral border, oxidative enzymes are localized.

DU 1 is characterized by the layering of the ganglion cells. The retina is now connected to the differentiating pigmented epithelium by gap junctions. The main bulk of cells in the retina is now arranged in vitreo-sclerally oriented columns; this is probably due to the appearance of Müller cell processes. In the region where the plexiform layers are to appear, $^H3$-glucosamine is incorporated. From DU 2 on, the surface area and the thickness of the retina increase. Gap junctions are still frequent, but their center of frequency is shifted from the dorsal to the ventral part of the retina. The maximum in the number of gap junctions is attained at the end of DU 2. Diaphorase is most intense in the innermost zone. In DU 3, a thin inner plexiform layer appears, where desmosomes and punctate junctions are found. A layering of the amacrine cells is indicated, and later in this DU the outer plexiform layer appears. Gap junctions are only preserved between the retina and the pigmented epithelium, not within the retina. But within this DU, they disappear here, too.

In DU 4, rod and cone nuclei can be discerned. All layers of the adult retina can be recognized at the end of this stage. DU 5 is mainly characterized by the appearance of many pycnotic nuclei within the nuclear layer. Enhanced cell death leads to a reduction of thickness of this layer up to DU 7. Puncta adhaerentia appear also in the outer nuclear layer. In DU 6, diaphorase shows an overall increase, and protein has a peak and is subsequently reduced. The thickness of the inner plexiform layer increases very much, and the outer plexiform layer attains its maximal radial extension.

5.3 Specific Differentiation

DU 1 is characterized by the occurrence of neurites in the optic tract, and in DU 2 these are bundled. Up to DU 3, they have established topographically organized connections to the optic center. After the production of a first rate, the number of synapses increases only slowly in DU 4, but in DU 5, synapses in the inner plexiform layer are formed at an enhanced rate, as applies to the synaptic vesicles in the axon terminals.

In DU 6, a greater amount of thick neurites contributes to the optic tract, and in DU 7 first myelinated fibers and first phagosomes appear. In DU 8, a maximum of α-bungarotoxin binding sites is attained, which is lowered thereafter.

5.4 Neurochemical Differentiation

GABA is the first transmitter to occur in the retina; it is found as early as DU 1. In DU 3 it increases along with glutamine and glycine. Acetylcholine is synthesized in detectable amounts, and 5-OH-tryptophan-decarboxylase has a specially high level of activity. Neuramine acid, phospholipids, and several gangliosids appear in DU 4. The level of GABA rises rapidly, and there are two systems that take up GABA, a high and a low affinity one. An increase in 5-HT, monoaminoxidase, and several specific enzymes is noted.

In DU 5, the phase of rapid GABA synthesis comes to an end. Acetylcholinesterase is found in the outer plexiform layer. The amount of visual pigment begins to rise. In DU 6, it rises hyperbolically, and AChE is also found in the inner plexiform layer. The ACh content is subject to changes in DU 7. Visual pigment content has attained half its maximum. The gangliosides and phospholipids have another peak. In DU 8, when MAO has a phase of highest activity GABA rises rapidly again. In DU 10, visual pigment attains a level which will not be changed. There is no further change in GABA concentration, but there is only one system taking up GABA. The activity of MAO is lowered to adult level. Gangliosides and phospholipids have a third peak.

5.5 Differentiation of Specific Activity

As soon as DU 3, a small positive potential is found after illumination, but this cannot represent receptor functioning. In DU 4, electrical stimulation of the tractus opticus evokes a potential in the optic center, which is later subject to certain changes.

The first visually evoked potentials in the optic center are found in DU 5, and the sensitivity to visual stimuli will increase now. With the onset of DU 6, the first detectable, immature ERG is found.

The a- and mainly the b-wave increase in size in DU 7. In DU 8, a1- and a2-waves can be discerned. At the end of DU 8, the ERG is adultlike in its form. A response to light flashes and to synchronous flicker light can be elicited. The amplitude of the ERG increases all the time. Sustained stimulation evokes on and off responses in the retina and the optic center. After DU 10, the ERG, especially the c-components increase to adult size and further units appear.

5.6 Miscellaneous

Prior to DU 1, the polarity of the retina is fixed and as early as this, morphogenetic cell death begins. DU 1 marks the point after which a transdifferentiation from pigment cells into retina cells is no longer possible. The retinotectal specification is attained in DU 2. In contrast to what had been found in DU 1, cultivated retina cells are transmutable into pigment cells in DU 3.

6 Summary

In the first part of this survey the development of the retina is treated, comparing the events studied in the fish *Tilapia*, in the amphibian *Xenopus*, in the chick, in the mouse and the rat, and in man. The temporal flow of the development can be subdivided into the phases of: (a) proliferation, characterized by a high rate of mitoses which later declines or ceases; (b) forming cell shapes and arising cellular and cell free layers; this phase begins very early and always proceeds in a vitreo-sclerad direction; (c) general cell differentiation, starting in neuroblasts which are interconnected by different types of junctions; this phase includes all kinds of changes which are not directly connected to specific retinal activity, as well as cell death; (d) specific differentiation, including the production of neurites and dendrites, of specific connections to the optic brain centers, and all those events mentioned in the second part.

These phases overlap each other and are more defined by their contents than by temporal boundaries. But proliferation is always the beginning of retina development; its end is always the appearance of specific physiological activity. The concept of a vitreo-sclerad sequence of differentiation is based upon light microscopic data, but on closer inspection, bipolar cells, not receptor cells, prove to be the last ones to differentiate. If those structures are looked at which are important for specific activity, one finds a vitreo-sclerad tendency, but after the onset in ganglion cells interneurons and receptor cells differentiate synchronously.

In the second part, the differentiation of those structures which are important for specific retina activity is reviewed.

The receptor cell inner segment enlarges and by formation of mitochondria, rER, and glycogen becomes characterized as a metabolic center; subsurface cisterns in double cones and oil droplets appear as specific structures.

Receptor cell outer segments consist of membrane disks formed by invagination of the plasma membrane of the outgrowing centriole. The growth of the outer segment is brought about by controlled stacking up of these disks. The visual pigment, combined of opsin from the inner segment and retinol from the pigmented epithelium, is incorporated into newly formed as well as already existing disks.

The receptor cell terminal synapses are formed before or after invagination of dendrites; usually, two horizontal cell dendrites and one central bipolar cell process, appearing somewhat later form a triadic structure. Receptor terminal synapses are ribbon synapses; the ribbon is of uncertain origin and shows an increase in length after its first appearance at the synaptic site.

The inner plexiform layer is the synaptic layer which takes its origin from outgrowing ganglion cell dendrites and amacrine and bipolar cell neurites, the latter again appearing as the last ones. Synaptic junctions are formed at certain contact regions throughout the layer, known as synaptic bands. Maturation of synaptic junctions is visualized by membrane asymmetry, pre- and postsynaptic densities, cleft densities, and synaptic vesicles. It is held as a rule that amacrine cells form conventional synapses, while bipolar cells form ribbon synapses opposite to a dyad of amacrine or ganglion cell dendrites, or both. Bipolar cell synapses are only functional when ribbons are seen; this is later than the maturation of conventional synapses. The development of synapse numbers consists of a rapid phase followed by a slow phase, or vice versa. Both synapse types develop at different rates, which probably has consequences in varying qualities of visual perception.

Among the transmitter systems, the cholinergic one is the best studied. In the chick there is a gradual increase of its single components as well as of acetylcholine receptors with a peak at about hatching. In the complex "synaptic junctions – transmitter components" both presynaptic and postsynaptic sites are involved and require some kind of coordination. The increases found in components of the cholinergic, the aminergic, and the GABAergic system are due to an increase in maturing cells of the whole retina. The appearance of response characteristics on the level of ganglion cells is temporally correlated to the achievement of definite receptive field size and dendritic morphology. The electroretinogram as representative of the activity of the retina reveals a differential appearance of components and changes in amplitude and sensitivity. Mostly, an a-component is the first one to appear, followed by a positive b-wave; this is rarely different. The a-wave is correlated to receptor cell development, the b-wave to Müller cells. A gradual increase in amplitude and a lowering of the sensitivity can also be attributed to an increase in number of mature cells.

Cultured embryonic retina tissue shows a high degree of self-differentiation; synapses, transmitter systems, and other typical elements are produced, with the exception of the outer segments.

Retina differentiation essentially occurs on two levels. Synapses, transmitter systems, outer segments and other elements which are needed for specific retina function mature within single cells. Retina maturity, however, as expressed, e.g., in the characteristics of the ERG, is acquired only by the multiplication of these single structures. Cell differentiation and multiplication – both are essential steps in tissue development.

Generally physiological maturity of the retina is achieved, when 60% of the time span between the onset and the end of overt differentiation have passed. [In extension of the data these conclusions are based upon, it is noted that Ali (1959) for the pacific salmon, and Armstrong (1964) for *Ictalurus*, reported light sensitivity and photic response in not fully developed retinae. Vogel (1978b) comes to the conclusion that in the cat retina final maturity is attained long after the onset of functioning.] The remaining 40% of the time span sees the formation of structures and components which are of secondary importance; mainly then, however, the multiplication of already existing structures occurs.

Synchronously, the more peripheral parts of the retina acquire maturity. There is, however, no study on the temporal relationship between central and peripheral parts of the retina, and of the question whether the peripherally directed differentiation proceeds steadily or stepwise. An interesting investigation by Dickson and Collard (1979) has shown that in the ammocoetes larva of a cyclostome, there is an abrupt separation of a well-differentiated central one-third and a totally undifferentiated peripheral zone.

68

The development of the retina is embedded into the general ontogenetic development, which in itself is of varying length in the different species. Computed into DUs, the onset of embryonic development in *Tilapia* takes place in DU −2. Its end, of course, is not easily fixed and for all species it is assumed to occur in DU 10. Thus in *Tilapia*, the total length of development is 12 DUs. Retina differentiation lasts from DU 1 to DU 8, covering 66% of the ontogenetic period. For *Xenopus*, the data are alike. In *Gallus* and *Homo*, the onset is at DU −1, and the retinal differentiation covers 73% of the altogether 11 DUs. The longest development is found in the rat, where it begins at DU −5. The retinal fraction is 53%. In the mouse it begins at DU −3, and 61% of the 13 DUs see the retinal development. Thus, the average retinal development period covers 65% or two-thirds of the embryonic development, beginning when between 9% and 33% of the latter have passed.

Apart from the obvious close relationship of the two rodents and certain similarities between *Tilapia* and *Xenopus*, attempts to reveal phylogenetic interrelations have remained without result. Apparently, the development of a retina has evolved in rather early ancestors and is a common heritage to all classes of vertebrates. The lack of greater phylogenetic discrepancies in adult retinal structure and the similarity of the cyclostome retina confirm this assumption and show that with the retina a very successful functional entity has evolved very early.

References

Ahlbert I-B (1973) Ontogeny of double cones in the retina of perch fry (Perca fluviatilis, Teleostei). Acta Zool (Stockh) 54:241–254

Ali MA (1959) The ocular structure, retinomotor and photobehavioral responses of juvenile pacific salmon. Can J Zool 37:965–996

Amemiya T, Ueno S (1977) Electron histochemical and developmental study on glycogen metabolism in retina of chick fetus. Acta Histochem 58:269–274

Anderson DH, Fisher SK, Steinberg RH (1978) Mammalian cones: Disc shedding, phagocytosis, and renewal. Invest Ophthalmol Vis Sci 17:117–133

Armengol JA, Prada F, Genis-Galvez JM (1979) Vesiculas cubiertas (coated vesicles) en la retina del embrion de pollo: Un primer estadio sinaptogenico? Morfol Norm Patol [A]3:707–713

Armstrong PB (1964) Photic responses in developing Bullhead embryos. J Comp Neurol 123:147–160

Babuchin A (1863) Beiträge zur Entwicklungsgeschichte des Auges, besonders der Retina. Würzbg Naturwiss Z 4:71–90

Baburina EA (1956) Development and function of the eyes in the sturgeon and the sterlet. (In Russian) Dokl Akad Nauk SSSR 106:359–361

Baburina EA (1972) Development of the eye in cyclostomes and fishes, in relation to ecology. (In Russian) Izdatielstvo "Nauka", Moscow

Baburina EA, Mitashov VI, Sinitzina VF, Lobacheva VA (1977) Appearance and transformation of the region of early photoreception in the eyes of the sturgeon embryos and prelarvae. An autoradiographic assay. (In Russian) Ontogenez 8:468–477

Bader CR, Baughman RW, Moore JL (1978) Different time course of development for high-affinity choline uptake and choline acetyltransferase in chick retina. Proc Natl Acad Sci USA 75:2525–2530

Baker PC, Quay WB (1969) 5-Hydroxytryptamine metabolism in early embryogenesis, and the development of brain and retinal tissues. A review. Brain Res 12:273–295

Barber AN (1955) Embryology of the human eye. Mosby, St Louis

Barnes SN (1974) Fine structure of the photoreceptor of the ascidian tadpole during development. Cell Tissue Res 155:27–46

Barr-Nea L, Barishak RY (1970) Tissue culture studies of the embryonal chicken retina. Invest Ophthalmol 9:447–457

Beach DH, Jacobson M (1979a) Patterns of cell proliferation in the retina of the clawed frog during development. J Comp Neurol 183:603–614

Beach DH, Jacobson M (1979b) Influences of thyroxine on cell proliferation in the retina of the clawed frog of different ages. J Comp Neurol 183:615–624

Berger ER (1964) Mitochondria genesis in the retinal photoreceptor inner segment. J Ultrastruct Res 11:90–111

Berger ER (1967) Subsurface cisterns in paired cone photoreceptor inner segments of adult and neonatal Lebistes retinae. J Ultrastruct Res 17:220–232

Berkow JW, Patz A (1964) Developmental histochemistry of the rat eye. Invest Ophthalmol 3:22–33

Berman AL, Azimova AM, Gribakin FG (1977) Localization of Na^+-, K^+-ATPase and Ca^{2+}-activated Mg^{2+}-dependent ATPase in retinal rods. Vision Res 17:527–536

Bhattacharjee J (1977) Sequential differentiation of retinal cells in the mouse studied by diaphorase staining. J Anat 123:273–282

Bhattacharjee J, Sanyal S (1975) Developmental changes of esterases in the retina of the mouse: Histochemical study. Histochemistry 46:53–60

Blanks JC, Bok D (1977) An autoradiographic analysis of postnatal cell proliferation in the normal and degenerative mouse retina. J Comp Neurol 174:317–329

Blanks JC, Adinolfi AM, Lolley RN (1974a) Synaptogenesis in the photoreceptor terminal of the mouse retina. J Comp Neurol 156:81–94

Blanks JC, Adinolfi AM, Lolley RN (1974b) Photoreceptor degeneration and synaptogenesis in retinal-degenerative (rd) mice. J Comp Neurol 156:95–106

Blaxter JHS, Jones MP (1967) The development of the retina and retinomotor responses in the herring. J Mar Biol Assoc UK 47:677–697

Blaxter JHS, Staines M (1970) Pure-cone retinae and retinomotor responses in larval teleosts. J Mar Biol Assoc UK 50:449–460

Blozovski D (1971a) Maturation des réponses visuelles évoquées par la stimulation électrique du nerf optique chez l'embryon de Poulet. J Physiol (Paris) 63:473–496

Blozovski D (1971b) Ontogénèse de la voie visuelle. Bull Psychol 291:9–11

Blozovski D, Blozovski M (1968) Développement composé de l'électrorétinogramme et des potentiels évoqués visuels du toit optique, du cervelet et du télencéphale chez le poussin. J Physiol (Paris) 60:33–50

Boell EJ, Greenfield P, Shen SC (1955) Development of cholinesterase in the optic lobes of the frog. J Exp Zool 129:415–451

Bok D (1968) An electron microscopic analysis of migration, division, and differentiation of presumptive rat photoreceptors. Anat Rec 160:Abstr 319

Bok D, Hall MO, Brien PO (1977) The biosynthesis of rhodopsin as studied by membrane renewal in rod outer segments. In: Brinkley BR, Porter KR (eds) International cell biology 1976–1977. Rockefeller University Press, New York, pp 608–617

Bonaventure NS, Goswamy S, Karli P (1967) Maturation des potentiels ERG et évoqués visuels chez le lapin élevé dans des conditions naturelles d'éclairement ambiant. CR Soc Biol (Paris) 161:689–693

Bonting SL, Caravaggio LL, Gouras P (1961) The rhodopsin cycle in the developing vertebrate retina. I. Relation of rhodopsin content, electroretinogram, and rod structure in the rat. Exp Eye Res 1:14–24

Bonting SL, Caravaggio LL, Hawkins NM (1962) Studies on sodium-potassium activated adenosine triphosphatase. IV. Correlation with cation transport sensitive to cardiac glycosides. Arch Biochem Biophys 98:413–419

Bornschein H (1959) Zur postnatalen Entwicklung der Netzhautfunktion. Vergleichende elektroretinographische Untersuchungen. Wien Klin Wochenschr 71:956–958

Bowe-Anders C, Miller RF, Dacheux R (1975) Developmental characteristics of receptive organization in the isolated retina-eye-cup of the rabbit. Brain Res 87:61–66

Bownds D (1967) Site of attachment of retinal in rhodopsin. Nature 216:1178–1181

Braekevelt CR, Hollenberg MJ (1970) Development of the retina of the albino rat. Am J Anat 127:281–302

Bridges CDB (1972) The rhodopsin-porphyropsin visual system. In: Dartnall HJA (ed) Handbook of sensory physiology, vol 7/1, Photochemistry of vision. Springer, Berlin Heidelberg New York, pp 417–480

Bridges CDB, Fong S-L (1979) Different receptors for distribution of peanut and ricin agglutinins between inner and outer segments of rod cells. Nature 282:513–515

Bridges CDB, Hollyfield JG, Witkovsky P, Gallin E (1977) The visual pigment and vitamin A of Xenopus laevis embryos, larvae, and adults. Exp Eye Res 24:7–14

Brockhoff V (1957) Zur Entwicklung der Sehzellen und zur Frage der Kernsekretion in der Retina. Anat Anz [Suppl]104:162–166

Buskirk DR, Thiery J-P, Rutishauser U, Edelman GM (1980) Antibodies to a neural cell adhesion molecule disrupt histogenesis in cultured chick retina. Nature 285:488–490

Cafferata R, Finston D, Nowak D (1979) A change in protein synthesis correlated with the time of retinal specification in Xenopus laevis. Dev Neurosci 2:1–6

Caley DW, Johnson C, Liebelt A (1972) The postnatal development of the retina in the normal and rodless CBA mouse. A light and electron microscopic study. Am J Anat 133:179–212

Carasso N (1958) Ultrastructure des cellules visuelles de larves d'amphibiens. C R Acad Sci [D] (Paris) 247:527–531

Carasso N (1959) Etude au microscope électronique de la morphogénèse du segment externe des cellules visuelles chez le Pleurodèle. C R Acad Sci [D] (Paris) 248:3058–3060

Carasso N (1960) Rôle de l'ergastoplasme dans l'élaboration du glycogène au cours de la formation du paraboloïde des cellules visuelles. C R Acad Sci [D] (Paris) 250:600–602

Cardinali DP, Rosner JM (1971) Retinal localization of the hydroxyindole-o-methyltransferase (HIOMT) in the rat. Endocrinology 89:301–303

Carter-Dawson LD, LaVail MM (1979) Rods and cones in the mouse retina. II. Autoradiographic analysis of cell generation using tritiated thymidine. J Comp Neurol 188:263–272

Chader GJ (1971) Hormonal effects on the neural retina. I. Glutamine synthetase development in the retina and liver of the normal and triiodthyronine treated rat. Arch Biochem Biophys 144:657–662

Chader GJ, Fletcher RT, Newsome DA (1974) Development of phosphodiesterase activity in chick retina. Dev Biol 40:378–380

Changeux P, Danchin A (1976) Selective stabilization of developing synapses as a mechanism for the specification of neural networks. Nature 264:705–712

Chen F, Witkovsky P (1978) Formation of photoreceptor synapses in retina of larval Xenopus. J Neurocytol 7:721–740

Chievitz JH (1887) Die Area und Fovea centralis beim menschlichen Foetus. Int Monatsschr Anat Physiol B IV

Chung SH, Stirling RV, Gaze RM (1975) Structural and functional development of retina in larval Xenopus. J Embryol Exp Morphol 33:915–941

Cima C, Grant P (1980) Ontogeny of the retina and optic nerve of Xenopus laevis. IV. Ultrastructural evidence of early ganglion cell differentiation. Dev Biol 76:229–237

Clayton RM, De Pomerai D, Pritchard DJ (1977) Experimental manipulation of alternative pathways of differentiation in cultures of embryonic chick neural retina. Dev Growth Differ 19:319–328

Cohen LH, Noell WK (1960) Glucose catabolism of rabbit retina before and after development of visual function. J Neurochem 5:253–276

Cooper TG, Meyer DB (1968) Ontogeny of retinal oil droplets in the chick embryo. Exp Eye Res 7:434–442

Coulombre AJ (1955) Correlations of structural and biochemical changes in the developing retina of the chick. Am J Anat 96:153–190

Coulombre AJ (1961) Cytology of the developing eye. Int Rev Cytol 11:161–194

Coulombre JL, Coulombre AJ (1965) Regeneration of neural retina from the pigmented epithelium in the chick embryo. Dev Biol 12:79–92

Cragg BG (1975) The development of synapses in the visual system of the cat. J Comp Neurol 160:147–167

Crescitelli F (1966) Electroretinogram of the frog during embryonic development. Science 151:1545–1547

Crescitelli F (1973) The visual pigment system of Xenopus laevis: Tadpoles and adults. Vision Res 13:855–865

Crisanti-Combes P, Privat A, Pessac B, Calothy G (1977) Differentiation of chick embryo neuroretina cells in monolayer cultures. An ultrastructural study. I. Seven-day retina. Cell Tissue Res 185:159–175

Crisanti-Combes P, Pessac B, Calothy G (1978) Choline acetyltransferase activity in chick embryo neuroretinas during development in ovo and in monolayer cultures. Dev Biol 65:228–233

Crossland WJ, Currie JR, Rogers LA, Cowan WM (1974) Evidence for a rapid phase of axoplasmic transport at early stages in the development of the visual system of the chick and frog. Brain Res 78:483–489

De Mello FG, Bachrach U, Nirenberg M (1976) Ornithine and glutamic acid decarboxylase activities in the developing chick retina. J Neurochem 27:847–851

De Mello FG (1978) The ontogeny of dopamine-dependent increase of adenosine 3′, 5′-cyclic monophosphate in the chick retina. J Neurochem 31:1049–1054

Demirchoglian GG, Mirzoian VS (1953) Development of electric reaction of the retina in ontogenesis. (In Russian) Dokl Akad Nauk SSSR 90:371–374

Denham S (1967) A cell proliferation study in the neural retina in the two-day rat. J Embryol Exp Morphol 18:53–66

De Pomerai DI, Clayton RM (1978) Influence of embryonic stage on the transdifferentiation of chick neural retina cells in culture. J Embryol Exp Morphol 47:179–193

De Robertis E (1956) Morphogenesis of the retinal rods. An electron microscope study. J Biophys Biochem Cytol [Suppl]2:209–218

De Robertis E, Lasansky A (1961) Ultrastructure and chemical organization of photoreceptors. In: Smelser GK (ed) The structure of the eye. Academic Press, New York

Detwiler SR, Laurens H (1921) Studies on the retina: Histogenesis of the visual cells in Amblystoma. J Comp Neurol 33:493–598

Detwiler SR (1932) Experimental observations upon the developing rat retina. J Comp Neurol 55: 473–492

Dickson DH, Collard TR (1979) Retinal development in the lamprey (Petromyzon marinus). Premetamorphic ammocoete eye. Am J Anat 154:321–336

Ditto M (1975) A difference between developing rods and cones in the formation of outer segment membranes. Vision Res 15:535–536

Dixon JS, Cronly-Dillon JR (1972) The fine structure of the developing retina in Xenopus laevis. J Embryol Exp Morphol 28:659–666

Dobson V, Teller DY (1978) Visual acuity in human infants: A review and comparison of behavioral and electrophysiological studies. Vision Res 18:1469–1485

Dowling JE, Boycott BB (1967) Organization of the primate retina: Electron microscopy. Proc R Soc Lond [Biol] 166:80–111

Dowling JE, Gibbons JR (1961) The effect of vitamin A deficiency on the fine structure of the retina. In: Smelser GK (ed) The structure of the eye. Academic Press, New York, pp 85–99

Dowling JE, Sidman RL (1962) Inherited retinal dystrophy in the rat. J Cell Biol 14:73–109

Dreyfus H, Urban PF, Edel-Harth S, Mandel P (1975) Developmental patterns of gangliosides and of phospholipids in chick retina and brain. J Neurochem 25:245–252

Dreyfus H, Edel-Harth S, Urban PF, Neskovic N, Mandel P (1977) Enzymatic synthesis of lactosylceramide by a galactosyltransferase from developing chicken retina. Exp Eye Res 25:1–7

Duke-Elder SS, Cook C (1963) System of Ophthalmology, vol. III, Normal and abnormal development, part I, embryology. Mosby, St Louis

Eakin RM (1964) The effect of vitamin A deficiency on photoreceptors in the lizard Sceloporus occidentalis. Vision Res 4:17–22

Eakin RM (1965) Differentiation of rods and cones in total darkness. J Cell Biol 25:162–165

Eakin RM, Westfall JA (1961) The development of photoreceptors in the stirnorgan of the tree-frog Hyla regilla. Embryology (Nagoya) 6:84–98

Easter SS, Johns PR, Baumann LR (1977) Growth of the adult goldfish eye. I. optics. Vision Res 17:469–479

Edds MV Jr, Gaze RM, Schneider GE, Irwin LN (1979) Specificity and plasticity of retinotectal connections. Neurosci Res Program Bull 17:245–275

Eramisheva NV (1956) The development of the retina in groundling (in Russian). Dokl Akad Nauk SSSR 109:1219–1221

Farber DB, Lolley RN (1977) Influence of visual cell maturation or degeneration on cyclic AMP content of retinal neurons. J Neurochem 29:167–170

Feeney L (1973) The interphotoreceptor space. I. Postnatal ontogeny in mice and rats. Dev Biol 32:101–115

Fisher LJ (1972) Changes during maturation and metamorphosis in the synaptic organization of the tadpole retina inner plexiform layer. Nature 235:392–393

Fisher LJ (1976) Synaptic arrays of the inner plexiform layer in the developing retina of Xenopus. Dev Biol 50:402–412

Fisher LJ (1979) Development of synaptic arrays in the inner plexiform layer of neonatal mouse retina. J Comp Neurol 187:359–373

Fisher LJ, Easter SS (1979) Retinal synaptic arrays: continuing development in adult goldfish. J Comp Neurol 185:373–380

Fisher S, Jacobson M (1970) Ultrastructural changes during early development of retinal ganglion cells in Xenopus. Z Zellforsch 104:165–177

Fisher SK, Linberg KA (1975) Intercellular junctions in the early human embryonic retina. J Ultrastruct Res 51:69–78

Fletcher RT, Chader GJ (1978) Cyclic nucleotides and protein kinase systems in the developing chick retina and pigment epithelium. Biochim Biophys Acta 544:45–52

Foerster H, Lierse W (1975) The vulnerability of early postnatal differentiation processes of the retina and the teratogenic effect of cyclophosphamide (Endoxan). Acta Anat 93:161–170

Fujisawa H, Nakamura H, Chin M (1974) The fine structure of the reconstructed neural retina of chick embryos. J Embryol Exp Morphol 31:139–149

Fujisawa H, Morioka H, Watanabe K, Nakamura H (1976a) A decay of gap junctions in association with cell differentiation of neural retina in chick embryonic development. J Cell Sci 22:585–597

Fujisawa H, Morioka H, Nakamura H, Watanabe K (1976b) Gap junctions in the differentiated neural retina of newly hatched chicks. J Cell Sci 22:597–606

Fujita S, Horii M (1963) Analysis of cytogenesis in chick retina by tritiated thymidine autoradiography. Arch Histol Jpn 23:359–365

Fulton AB, Craft JI, Albert DM (1978) Human retinal dysplasia. Invest Ophthalmol Visual Sci [Suppl] 8:116

Galbavy ESJ, Olson MD (1978) Fine structure of developing photoreceptors in the postnatal rat. Anat Rec 190:398

Garcia-Austt E, Patetta-Queirolo MA (1961) Electroretinogram of the chick embryo. I. Onset and development. Acta Neurol Lat Am 7:179–189

Garcia-Porrero JA, Ojeda JL (1979) Cell death and phagocytosis in the neuroepithelium of the developing retina – TEM and SEM study. Experientia 35:375–377

Gaze RM, Peters A (1961) Development, structure, and composition of the optic nerve of Xenopus laevis (Daudin). Q J Exp Physiol 46:229–309

Goldberg S (1974) Studies on the mechanism of development of the visual pathways in the chick embryo. Dev Biol 36:24–44

Goldberg S (1976) Progressive fixation of morphological polarity in the developing retina. Dev Biol 53:126–128

Goldberg S (1977) Unidirectional, bidirectional, and random growth of embryonic optic axons. Exp Eye Res 25:399–405

Goldberg S, Coulombre AJ (1972) Topographical development of the ganglion cell fiber layer in the chick retina. A whole mount study. J Comp Neurol 146:507–519

Goto M (1950) The ontogenetic study of electroretinogram of chick embryo. (In Japanese) J Physiol Soc Jpn 12:67–71

Govardovsky VJ, Kharkeevich TA (1965) Histochemical and electron microscopical study of photoreceptive cell development under conditions of tissue culture. (In Russian) Arkh Anat Gistol Embriol 49:50–56

Govardovsky VJ, Kharkeevich TA (1966) Embryological development of photoreceptors. In: Academy of Sciences USSR (ed) Nervous processes in receptor elements of sense organs. (In Russian) Moscow, Leningrad, pp 4–22

Graham LT Jr (1974) Comparative aspects of neurotransmitters in the retina. In: Davson H, Graham LT (eds) The eye, vol 6. Academic Press, New York, pp 283–342

Graymore C (1959) Metabolism of the developing retina. I. Aerobic and anaerobic glycolysis in the developing rat retina. Br J Ophthalmol 43:34–39

Graymore C (1960) Metabolism of the developing retina. II. Respiration in the developing normal rat retina, and the effect of an inherited degeneration of the retina neuroepithelium. Br J Ophthalmol 44:363–369

Grillo MA, Rosenbluth J (1972) Ultrastructure of developing Xenopus retina before and after ganglion cell specification. J Comp Neurol 145:131–141

Grün G (1974) Elektronenmikroskopische Untersuchung zur Differenzierung der Rezeptoraußenglieder in der Retina von Tilapia leucosticta (Cichlidae). Verh Dtsch Zool Ges. Fischer, Stuttgart, pp 167–170

Grün G (1975) Structural basis of the functional development of the retina in the cichlid Tilapia leucosticta. J Embryol Exp Morphol 33:243–257

Grün G (1977a) Cytochemical localization of adenosine triphosphatase in the developing retina of a teleost. Vision Res 17:875–879

Grün G (1977b) The ultrastructural differentiation of synaptic sites in the inner plexiform layer of a teleostan retina. Z Mikrosk Anat Forsch 91:687–703

Grün G (1979) Light-induced acceleration of retina development in a mouth-brooding teleost. J Exp Zool 208:291–302

Grün G (1980) Developmental dynamic in synaptic ribbons of retinal receptor cells (Tilapia, Xenopus). Cell Tissue Res 207:331–339

Hamburger V, Hamilton HL (1951) A series of normal stages in the development of the chick embryo. J Morphol 88:49–92

Hanawa J, Takahashi K, Kawamoto N (1976) A correlation of embryogenesis of visual cells and early receptor potential in the developing retina. Exp Eye Res 23:587–594

Hasama B (1941) Ob und in welcher Embryonalzeit wird die Netzhaut des Huhns für verschiedene Strahlen empfindlich? Pflügers Arch 244:337–346

Hayes BP (1976) The distribution of intercellular gap junctions in the developing retina and pigment epithelium of Xenopus laevis. Anat Embryol 150:99–112

Hayes BP (1977) Intercellular gap junctions in the developing retina and pigment epithelium of the chick. Anat Embryol 151:325–334

Heacock AM, Agranoff BW (1977) Clockwise growth of neurites from retinal explants. Science 198:64–67

Heaton MB (1973) Early visual function in bobwhite and japanese quail embryos as reflected by pupillary reflex. J Comp Phys Psychol 84:134–139

Heaton MB, Munson JB (1978) Visual system development in the duck embryo. Exp Neurol 59: 53–61

Hebel R (1971) Entwicklung und Struktur der Retina und des Tapetum lucidum des Hundes. Ergeb Anat Entwicklungsgesch 45:1–93

Hellner KA, Utermann D (1965) Quantitative elektroretinographische Untersuchungen über die Entwicklung der normalen und dystrophischen Rattennetzhaut. Vision Res 5:535–544

Hild W, Callas G (1967) The behavior of retinal tissue in vitro, light and electron microscopic observations. Z Zellforsch 80:1–21

Himstedt W (1970) Das Elektroretinogramm des Feuersalamanders (Salamandra salamandra L.) vor und nach der Metamorphose. Pflügers Arch 318:176–184

Hinds JW, Hinds PL (1974) Early ganglion cell differentiation in the mouse retina. Dev Biol 37: 381–416

Hinds JW, Hinds PL (1976) An EM serial section study of the mouse retina at the 15th day of gestation. Anat Rec 184:A428

Hinds JW, Hinds PL (1978) Early development of amacrine cells in the mouse retina – an electron microscopical serial section analysis. J Comp Neurol 179:277–300

Hollenberg MJ, Spira AW (1973) Human retinal development: Ultrastructure of the outer retina. Am J Anat 137:357–387

Hollyfield JG (1972) Histogenesis of the retina in the killifish, Fundulus heteroclitus. J Comp Neurol 144:373–380

Hollyfield JG, Rayborn ME (1979) Photoreceptor outer segment development – Light and dark regulate the rate of membrane addition and loss. Invest Ophthalmol Vis Sci 18:117–132

Hollyfield JG, Witkovsky P (1974) Pigmented retinal epithelium involvement in photoreceptor development and function. J Exp Zool 189:357–378

Hollyfield JG, Mottow LS, Ward A (1975) Autoradiographic study of (3H)glucosamine incorporation by the developing retina of the clawed toad, Xenopus laevis. Exp Eye Res 20:382–393

Hollyfield JG, Rayborn ME, Sarthy PV, Lam DMK (1979) The emergence, localization, and maturation, of neurotransmitter systems during development of the retina in Xenopus laevis. I. γ-amino-butyric acid. J Comp Neurol 188:587–598

Horsten GPM, Winkelman JE (1960) Development of the ERG in relation to histological differentiation of the retina in man and animals. Arch Ophthalmol 63:232–242

Horsten GPM, Winkelman JE (1962) Electrical activity of the retina in relation to histological differentiation in infants born prematurely and at full term. Vision Res 2:269–276

Hughes WF, Lavelle A (1974) On the synaptogenic sequence in the chick retina. Anat Rec 179: 297–301

Hughes WF, Lavelle A (1975) The effect of early tectal lesions on development in the retinal ganglion cell layer of chick embryos. J Comp Neurol 163:265–285

Hunt HH (1961) A study of the fine structure of the optic vesicle and lens placode of the chick embryo during induction. Dev Biol 3:175–209

Hunt WE, Goldring S (1951) Maturation of evoked responses of the visual cortex in the postnatal rabbit. Electroencephalogr Clin Neurophysiol 3:465–471

Jacobson M (1968a) Development of neuronal specificity in retinal ganglion cells of Xenopus. Dev Biol 17:202–218

Jacobson M (1968b) Cessation of DNA synthesis in retinal ganglion cells correlated with the time of specification of their central connections. Dev Biol 17:219–232

Jacobson M (1976a) Premature specification of the retina in embryonic Xenopus eyes treated with Ionophore X537A. Science 191:288–290

Jacobson M (1976b) Neural recognition in the retinotectal system. In: Barondes SH (ed) Neuronal recognition. Chapman & Hall, London, pp 3–23

Johns PR, Yoon MG, Agranoff BW (1978) Directed outgrowth of optic fibres regenerating in vitro. Nature 271:360–362

Kahn AJ (1973a) The order of appearance of neurons in the developing chick retina. ARVO Spring Meeting, Sarasota, Fla. Personal communication

Kahn AJ (1973) Ganglion cell formation in the chick neural retina. Brain Res 63:285–291

Kahn AJ (1974) An autoradiographic analysis of the time of appearance of neurons in the developing chick neural retina. Dev Biol 38:30–41

Karli P (1961) Epithélium pigmentaire et différentiation de l'article externe des bâtonnets rétiniens. C R Soc Biol (Paris) 155:1694–1697

Kataoka S (1955) Cytochemistry of the retina (37). Electrophoretic studies on tissue proteins of the retina and choroid of the embryonal eye (report I). Acta Soc Ophthalmol Jpn 59:1603–1606

Keefe JR, Ordy JM, Samurajski TS (1966) Prenatal development of the retina in a diurnal primate (Macaca mulatta). Anat Rec 154:759–783

Keeler CE, Sutcliff E, Chaffee E (1928) A description of the ontogenetic development of retinal action currents in the house mouse. Proc Natl Acad Sci USA 14:811–815

Kinney MS, Fisher SK (1978a) The photoreceptors and pigment epithelium of the adult Xenopus retina: morphology and outer segment renewal. Proc R Soc Lond [Biol] 201:131–147

Kinney MS, Fisher SK (1978b) The photoreceptors and pigment epithelium of larval Xenopus retina: morphogenesis and outer segment renewal. Proc R Soc Lond [Biol] 201:149–167

Kinney MS, Fisher SK (1978c) Changes in length and disk shedding rate of Xenopus rod outer segments associated with metamorphosis. Proc R Soc Lond [Biol] 201:169–177

Kolmer W (1936) Entwicklung des Auges. In: Möllendorff W v. (ed) Handbuch der mikroskopischen Anatomie des Menschen, vol 3/2. Springer, Berlin Heidelberg New York, pp 623–676

Kunz YW, Wise C (1974) Development of the photoreceptors in the embryonic retina of Lebistes reticulatus (Peters). Electron microscopical investigations. Rev Suisse Zool 81:697–701

Kuo ZY (1932) Ontogeny of embryonic behavior in Aves. I. The chronology and general nature of the behavior of the chick embryo. J Exp Zool 61:395–430

Kuwabara T, Weidman TA (1974) Development of the prenatal rat retina. Invest Ophthalmol 13:725–729

Lam DMK (1976) Synaptic chemistry of identified cells in the vertebrate retina. Cold Spring Harbor Symp Quant Biol 40:571–579

Lam DMK (1977) Electroretinogram of the newt during retinal regeneration. Brain Res 136:148–153

Lanzavecchia G (1959) Ultrastruttura dei coni nella retina dell'embrione di Xenopus laevis. In: Società italiana di microscopia elettronica (ed) Atti del IIo congresso italiano di microscopia elettronica, Milan, pp 33–38

LaVail MM (1973) Kinetics of rod outer segment renewal in the developing mouse retina. J Cell Biol 58:650–662

LaVail MM, Hild W (1971) Histotypic organization of the rat retina in vitro. Z Zellforsch 114:557–579

LaVail MM, Reif-Lehrer L (1971) Glutamine synthetase in the normal and dystrophic mouse retina. J Cell Biol 51:348–354

Lee SH, Pak SY, Choi KD (1967) A histochemical study of cholinesterase activity in rabbit's retina. Yonsei Med J 8:1–7

Leplat G (1914) Les plastosomes des cellules visuelles et leur rôle dans la différenciation des cones et des bâtonnets. Anat Anz 45:215–221

Lerche W (1963) Elektronenmikroskopische Untersuchungen zur Differenzierung des Pigmentepithels und der äußeren Körnerzellen (Sinneszellen) im menschlichen Auge. Z Zellforsch 58:953–970

Levick WR (1972) Receptive fields of retinal ganglion cells. In: Fuortes MGF (ed) Handbook of sensory physiology, vol VII/2. Springer, Berlin Heidelberg New York, pp 531–566

Levick WR (1975) Form and function of the retinal ganglion cells. Nature 254:695–663

Levine RL (1979) Involvement of retinal pigment epithelium in the genesis of locus specifities during retinal regeneration in the newt. Brain Res 162:154–158

Lewis MR (1922) Mitochondria in the visual cells of the fowl. Anat Rec 25:110–111

Linberg KA, Fisher SK (1978) Ultrastructural changes in the outer retina during cone synaptogenesis in humans. ARVO Spring Meeting Sarasota, Fla. Invest Ophthalmol Vis Sci [Suppl], p 116

Lindeman VF (1947) The cholinesterase and acetylcholine content of the chick retina with special reference to functional activity as indicated by the pupillary constrictor reflex. Am J Physiol 148:40–44

Lindeman VF (1949) Alkaline and acid phosphatase activity of the embryonic chick retina. Proc Soc Exp Biol Med 71:435–437

Lolley RN, Racz E (1972) Changes in levels of ATPase activity in developing retinae of normal (DBA) and mutant (C3H) mice. Vision Res 12:567–573

Lolley RN, Schmidt SY, Farber DB (1974) Alterations in cyclic AMP metabolism associated with photoreceptor cell degeneration in the C3H mouse. J Neurochem 22:701–707

Lucas DR (1961) The effect of X-radiation on the mouse retina at different stages of development. Int J Radiat Biol 3:105–124

Lucas DR (1965) Special cytology of the eye. In: Willmer EN (ed) Cells and tissues in culture, vol 2. Academic Press, London New York, pp 457–520

Ma PM, Grant P (1978) Ontogeny of ACh and GABA synthesis during development of the Xenopus retina. Brain Res 140:368–374

Macaione S, Cacioppo F (1971) Glutamic acid metabolism in rat retina during postnatal growth. Ital J Biochem 20:112–123

Macaione S, Campisi R, Albanese A (1970) Glutamato decarbossilasi e γ-aminobutirato transaminasi nella retina di ratto durante la crescita postnatale. Boll Soc Ital Biol Sper 46:1–4

Macaione S, Cacioppo F, Campisi R, Ruggeri P (1973) Transaminases in rat retina during development. Life Sci 13:1429–1434

Macaione S, Ruggeri P, Deluca F, Tucci G (1974) Free amino acids in developing rat retina. J Neurochem 22:887–892

Macri JJ, Rebollo MA (1965) Diferenciacion de la retina y la via optica en el embrion de pollo. IV. Enzimas en la retina durante el desarrollo. Acta Neurol Lat Am 11:309–320

Mann I (1928) The process of differentiation of the retinal layers in vertebrates. Br J Ophthalmol 12:449–479

Mann I (1964) The development of the human eye. Grane & Stratton, New York

Marchisio PC, Sjöstrand J, Aglietta M, Karlsson J-O (1973) The development of axonal transport of proteins and glycoproteins in the optic pathway of chick embryo. Brain Res 63:273–285

Martinozzi M, Moscona AA (1975) Binding of ^{125}J-concanavalin A and agglutination of embryonic neural retina cells. – Age dependent and experimental changes. Exp Cell Res 94:253–266

Masland RH (1977) Maturation of function in the developing rabbit retina. J Comp Neurol 175:275–286

Mason WT, Bighouse KJ (1975) Correlation of rhodopsin biogenesis with ultrastructural morphogenesis in the chick retina. J Cell Biol 64:235–241

Mason WT, Fager RS, Abrahamson EW (1973) Ultrastructure of the receptor and epithelial layers of the bovine retina. J Anat 115:289–308

Masterson E, Israel P, Chader GJ (1978) Pentose shunt activity in developing chick retina and pigment epithelium: A switch in biochemical expression in cultured pigment epithelium cells. Exp Eye Res 27:409–416

McArdle CB, Dowling JE, Masland RH (1977) Development of outer segments and synapses in the rabbit retina. J Comp Neurol 175:253–273

McLaughlin BJ (1976a) A fine structure and E-PTA study of photoreceptor synaptogenesis in the chick retina. J Comp Neurol 170:347–364

McLaughlin BJ (1976b) Photoreceptor synaptogenesis in the chick retina. Anat Rec 184:A 476

McLaughlin BJ, Wood JG (1977) The localization of concanavalin A binding sites during photoreceptor synaptogenesis in the chick retina. Brain Res 119:57–72

McLaughlin BJ, Wood JG, Gurd JW (1980) The localization of lectin binding sites during photo-receptor synaptogenesis in the chick retina. Brain Res 191:345–359

McLoon SC, Hughes WF (1978) Ganglion cell death during retinal development in chick eyes trans-planted to the chorioallantoic membrane. Brain Res 150:398–402

Meller K (1964) Elektronenmikroskopische Befunde zur Differenzierung der Rezeptorzellen und Bipolarzellen der Retina und ihrer synaptischen Verbindungen. Z Zellforsch 64:733–750

Meller K (1968) Histo- und Zytogenese der sich entwickelnden Retina: Eine elektronenmikro-skopische Studie. Fischer, Stuttgart

Meller K, Breipohl W (1965) Die Feinstruktur und Differenzierung des inneren Segmentes und des Paraboloids der Photorezeptoren in der Retina von Hühnerembryonen. Z Zellforsch 66:673–684

Meller K, Glees P (1965) The differentiation of neuroglia Müller cells in the retina of the chick. Z Zellforsch 66:321–332

Meller K, Tetzlaff W (1976) Scanning electron microscopic studies on the development of the chick retina. Cell Tissue Res 170:145–160

Meller K, Tetzlaff W (1977) The development of membrane specializations in the receptor – bipolar – horizontal cell synapse of the chick embryo retina. A freeze-fracture study. Cell Tissue Res 181:319–327

Meyer H (1936) Züchtung der Retina des Huhnes in vitro. Z Mikrosk Anat Forsch 39:151–160

Mikhailov AT, Gorgolyuk NA (1976) An electrophoretic study of water-soluble proteins of chick retina at the early embryonic stages (in Russian). Ontogenez 7:333–340

Mintz G, Glaser L (1978) Specific glycoprotein changes during development of the chick neural retina. J Cell Biol 79:132–137

Mishima H, Fujita H (1978) Studies on the cytodifferentiation of the neuroblasts and visual cell in the chick embryo retina, using the electron microscopic autoradiography of 3H-thymidine. Albrecht Von Graefes Arch Klin Exp Ophthalmol 206:1–10

Mock BH, Weidman TA, Carlton WW, Landolt RR, Shaw SM (1975) The effect of diagnostic-quality X-irradiation on the developing rat retina. Exp Eye Res 20:293–307

Monard D, Solomon F, Rentsch M, Gysin R (1973) Glia-induced morphological differentiation in neuroblastoma cells. Proc Natl Acad Sci USA 70:1894–1897

Morest DK (1970) The pattern of neurogenesis in the retina of the rat. Z Anat Entwicklungs-gesch 131:45–67

Morris JE, Hopwood JJ, Dorfman A (1977) Biosynthesis of glycosaminglycans in the developing retina. Dev Biol 58:313–327

Morris VB (1973) Time difference in the formation of the receptor types in the developing chick retina. J Comp Neurol 151:323–330

Morris VB (1974) Triple cones in the chick retina. In: 8th Int. Congress Electr. Micr., Canberra, vol II, p 312–313

Morris VB (1975) Non-randomness in the sequential formation of principle cones in small areas of the developing chick retina. J Comp Neurol 164:95–105

Moscona AA (1976) Cell recognition in embryonic morphogenesis and the problem of neuronal specification. In: Barordes SH (ed) Neuronal recognition. Chapman & Hall, London, pp 205–226

Müller H (1952) Bau und Wachstum der Netzhaut des Guppy (Lebistes reticulatus). Zool Jahrb Abt Allg Zool Physiol Tiere 63:275–324

Müller-Limmroth HW, Andrée G (1954) Die ontogenetische Entwicklung des Elektroretinogramms des Frosches. Z Biol 107:25–33

Munk O, Andersen SR (1962) Acessory outer segment, a rediscovered cilium-like structure in the layer of rods and cones in human retina. Acta Ophthalmol 40:526–531

Nakayama K (1957) Histochemical studies on the human fetal retina in the course of development (2 nucleic acids). J Clin Ophthalmol (Tokyo) 11:1024–1032

Nieuwkoop PD, Faber J (1967) Normal table of Xenopus laevis (Daudin). North-Holland, Amster-dam

Nilsson SEG (1964) Receptor outer segment development and ultrastructure of the disk membra-nes in the retina of the tadpole (Rana pipiens). J Ultrastruct Res 11:581–620

Nilsson SEG, Crescitelli F (1969) Changes in ultrastructure and electroretinogram of bullfrog retina during development. J Ultrastruct Res 27:45–62

Nilsson SEG, Crescitelli F (1970) A correlation of ultrastructure and function in the developing retina of the frog tadpole. J Ultrastruct Res 30:87–102

Nishimura Y (1980) Determination of the developmental pattern of retinal ganglion cells in chick embryos by Golgi impregnation and other methods. Anat Embryol 158:329–347

Noell WK (1958) Differentiation, metabolic organization and viability of the visual cell. Arch Ophthalmol 60:702–731

Norenberg MD, Dutt K, Reif-Lehrer L (1980) Glutamine synthetase localization in cortisol induced chick embryo retinas. J Cell Biol 84:803–807

Okada TS, Yasuda K, Araki M, Eguchi G (1979) Possible demonstration of multipotential nature of embryonic neural retina by clonal cell culture. Dev Biol 68:600–617

Olive J, Recouvreur M (1977) Differentiation of retinal rod disc membranes in mice. Exp Eye Res 25:63–74

Olney JW (1968a) An electron microscopic study of synapse formation, receptor outer segment development, and other aspects of developing mouse retina. Invest Ophthalmol 7:250–268

Olney JW (1968b) Centripetal sequence of appearance of receptor-bipolar synaptic structures in developing mouse retina. Nature 218:281–282

Olson MD (1972) Fine structural development of the chick retina: early morphogenesis of photoreceptors. Anat Rec 172:441

Olson MD (1973) Fine structural organization of photoreceptors in the chick: centrioles and ciliary development. Anat Rec 175:402

Olson MD (1979) Scanning electron microscopy of developing photoreceptors in the chick retina. Anat Rec 193:423–438

Ookawa T (1971a) The onset and development of the chick electroretinogram: The A- and B-waves. Poult Sci 50:601–608

Ookawa T (1971b) Further studies on the ontogenetic development of the chick electroretinogram. Poult Sci 50:1185–1190

Ookawa T, Takahashi K (1971) The ontogenetic development of the c-wave in the chick ERG. Experientia 27:407–409

Oppenheim RW (1968) Light responsivity in chick and duck embryos just prior to hatching. Anim Behav 16:276–280

O'Rahilly R (1975) The prenatal development of the human eye. Exp Eye Res 21:293–313

O'Rahilly R, Meyer DB (1955) Correlation between the development of the eye and embryonic staging in the chick. Anat Rec 121:346

O'Rahilly R, Meyer DB (1959) The early development of the eye in the chick Gallus domesticus (stage 8–25). Acta Anat 36:20–58

Ordy JM, Masopust LC, Wolin LR (1962) Postnatal development of the retina, electroretinogram, and acuity in the monkey. Exp Neurol 5:364–382

Ordy JM, Samorajski T, Collins RL, Nagy AR (1965) Postnatal development of vision in a sub-human primate (Macaca mulatta). A multidisciplinary study. Arch Ophthalmol 73:674–686

Orr HT, Cohen AI, Carter JA (1976) The levels of free taurine, glutamate, glycine, and γ-aminobutyric acid during the postnatal development of the normal and dystrophic retina of the mouse. Exp Eye Res 23:377–384

Orts-Llorca F, Genis-Galvez JM (1960) Experimental production of retinal septa in the chick embryo. Differentiation of pigment epithelium into neural retina. Acta Anat 42:31–70

Papermaster DS, Converse CA, Siu J (1975) Membrane biosynthesis in frog retina: Opsin transport in the photoreceptor cell. Biochemistry 14:1343–1352

Pasantes-Morales H, Klethi J, Ledig M, Mandel P (1973) Influence of light and dark on the free amino acid pattern of the developing chick retina. Brain Res 57:59–63

Pasantes-Morales H, Lopez-Colomé, Salceda R, Mandel P (1976) Cysteine sulphinate decarboxylase in chick and rat retina during development. J Neurochem 27:1103–1106

Paulson GW (1965) Maturation of evoked responses in the duckling. Exp Neurol 11:324–333

Peck D (1964) The role of tissue organization in the differentiation of embryonic chick neural retina. J Embryol Exp Morphol 12:381–390

Peters JJ, Vonderahe AR, Powers TH (1958) Electrical studies of functional development of the eye and optic lobes in the chick embryo. J Exp Zool 139:459–468

Pomeranz B (1972) Metamorphosis of frog vision: Changes in ganglion cell physiology and anatomy. Exp Neurol 34:187–199

Pomeranz B, Chung SH (1970) Dendritic-tree anatomy codes form-vision physiology in tadpole retina. Science 170:983–984

Raedler A, Sievers J (1975) The development of the visual system of the albino rat. Adv Anat Embryol 50:1–88

Rager G (1976) Morphogenesis and physiogenesis of the retinotectal connections in the chicken. I. The retinal ganglion cells and their axons. Proc R Soc Lond [Biol] 192:331–352

Rager G (1979) The cellular origin of the b-wave in the electroretinogram – A developmental approach. J Comp Neurol 225–244

Rager G, Rager U (1978) Systems-matching by degeneration. I. A quantitative electron microscopic study of the generation and degeneration of retinal ganglion cells in the chicken. Exp Brain Res 33:65–78

Ramirez G (1977) Cholinergic development in chick optic tectum and retina reaggregated in cell cultures. Neurochem Res 2:427–438

Raviola E, Raviola G (1962) Ricerche istochimiche sulla retina di coniglio nel corso dello sviluppo postnatale. Z Zellforsch 56:552–572

Reading HW (1965) Protein biosynthesis and the hexosemonophosphat shunt in the developing normal and dystrophic retina. In: Graymore CN (ed) Biochemistry of the retina. Academic Press, London New York, pp 73–82

Rebollo MA (1955) Some aspects of the histogenesis of the retina. Acta Neurol Lat Am 1:142–147

Rebollo MA (1963) Diferenciacion de la retina y la via optica en el embrion de pollo. I. Diferenciacion de las capas de la retina. Acta Neurol Lat Am 9:189–194

Reuter JH (1976) The development of the electroretinogram in normal and light-deprived rabbits. Pflügers Arch 363:7–13

Reuter T (1969) Visual pigments and ganglion cell activity in the retina of tadpoles and adult frogs (Rana temporaria L.). Acta Zool Fenn 122:1–64

Rhodes RH (1979) A light microscopic study of the developing human neural retina. Am J Anat 154:195–210

Rickenbacher J (1952) Die Nukleinsäuren in der Augenentwicklung bei Amphibien und beim Hühnchen. Wilhelm Roux' Arch Entwicklungsmech Org 145:387–402

Riepe RE, Norenberg MD (1978) Glutamine synthetase in the developing rat retina: an immunohistochemical study. Exp Eye Res 27:435–444

Robinson WE, Yoshikami S, Hagins WA (1975) ATP in retinal rods. Biophysical J 15:168a

Rodieck RW (1973) The vertebrate retina. Freeman, San Francisco

Rogers KT (1957) Early development of the optic nerve in the chick. Anat Rec 127:97–107

Rohrschneider I (1975) Licht- und elektronenmikroskopische Untersuchung an der sich entwickelnden Retina des Blauen Zwergmaulbrüters Haplochromis burtoni (Cichlidae, Teleostei). Verh Anat Ges 69:659–663

Rudnick D (1959) Glutamotransferase and histogenesis in the transplanted chick retina. J Exp Zool 142:643–666

Ruffolo RR, Eisenbarth GS, Thompson JM, Nirenberg M (1978) Synapse turnover – Mechanism for acquiring synapse specificity. Proc Natl Acad Sci USA, pp 2281–2286

Rusoff AC, Dubin MW (1977) Dévelopment of receptive field properties of retinal ganglion cells in kitten. J Neurophysiol 40:1188–1199

Rusoff AC, Dubin MW (1978) Kitten ganglion cells – dendritic field size at 3 weeks of age and correlation with receptive field size. Invest Ophthalmol Vis Sci 17:819–821

Sack RA, Harris CM (1977) Ca^{2+} dependent ATPase activity of bovine receptor cell outer segment. Nature 265:465–466

Samson-Dollfus D (1968) Developpement normal de l'électrorétinogramme depuis l'âge foetal de sept mois et demi jusqu'à l'âge de quatre mois après la naissance à terme. Bull Soc Ophthalmol Fr 4:423–431

Saxén L (1954) The development of the visual cells. Embryological and physiological investigations on amphibia. Ann Acad Sci Fenn [Biol] 23:1–93

Schmidt B (1973) Die Histotopographie der Phosphatasen während der Entwicklung der Rattenretina und ihre Störung durch Röntgenstrahlen. Acta Anat 84:161–177

Schook P (1978) A review on data of cell actions and cell interactions during the morphogenesis of the embryonic eye. Acta Morphol Neerl Scand 16:267–286

Seefelder R (1910) Beiträge zur Histogenese und Histologie der Netzhaut, des Pigmentepithels und des Sehnerven. Albrecht von Graefes Arch Ophthalmol 73:419–537

Sheffield JB (1970) Studies on aggregation of embryonic cells: Initial cell adhesions and the formation of intercellular junctions. J Morphol 132:245–264

Sheffield JB, Fischman DA (1970) Intercellular junctions in the developing neural retina of the chick embryo. Z Zellforsch 104:405–418

Sheffield JB, Moscona AA (1970) Electron microscopic analysis of aggregation of embryonic cells: The structure and differentiation of aggregates of neural retina cells. Dev Biol 23:36–61

Shen SC, Greenfield P, Boell EJ (1956) Localization of acetylcholinesterase in chick retina during histogenesis. J Comp Neurol 106:433–462

Shiragami M (1968) Electron microscopic study on synapses of visual cells. II. The morphogenesis of synapses of visual cells in chick embryo. Jpn J Ophthalmol 72:232–245

Sidman RL (1961a) Histogenesis of mouse retina studied with thymidine-3H. In: Smelser GK (ed) The structure of the eye. Academic Press, New York, pp 487–505

Sidman RL (1961b) Tissue culture studies of inherited retinal dystrophy. Dis Nerv Syst Monogr [Suppl] 32

Silver J (1976) A study of ocular morphogenesis in the rat using 3H-thymidine autoradiography: Evidence for thymidine recycling in the developing retina. Dev Biol 49:487–495

Silverstein AM, Osburn BJ, Prendergast RA (1971) The pathogenesis of retinal dysplasia. Am J Ophthalmol 72:13–26

Sjöstrand J, Karlsson J-O, Marchisio PC (1973) Axonal transport in growing and mature retinal ganglion cells. Brain Res 62:395–399

Smelser GK, Ozanics V, Rayborn M, Sagun D (1973) The fine structure of the retinal transient layer of Chievitz. Invest Ophthalmol 12:504–512

Smelser GK, Ozanics V, Rayborn M, Sagun D (1974) Retinal synaptogenesis in the primate. Invest Ophthalmol 13:340–362

Sorsby A, Koller PC, Attfield M, Davey JB, Lucas DR (1959) Retinal dystrophy in the mouse: Histological and genetic aspects. J Exp Zool 125:171–179

Spadaro A, de Simone I, Puzzolo D (1978) Ultrastructural data and chronobiological patterns of the synaptic ribbons in the outer plexiform layer in the retina of albino rats. Acta Anat 102: 365–373

Spira AW (1974) Ultrastructural localization of cholinesterase activity in the developing rat retina. J Histochem Cytochem 22:868–880

Spira AW (1975) In utero development and maturation of the retina of a non-primate mammal: A light and electron microscopic study of the guinea pig. Anat Embryol 146:279–300

Spira AW (1976) The localization of cholinesterase in the retina of the fetal and newborn guinea pig. J Comp Neurol 169:393–408

Spira AW, Hollenberg MJ (1973) Human retinal development: Ultrastructure of the inner retinal layers. Dev Biol 31:1–21

Spira AW, Huang PT (1978) Phagocytosis of photoreceptor outer segments during retinal development in utero. Am J Anat 152:523–528

Stefanelli A, Cataldi E, Ieradi LA (1961a) Istochimica della fosfatasi acida durante lo sviluppo embrionale della retina di pollo. Rend Accad Naz Lincei 8 30:664–667

Stefanelli A, Zacchei AM, Ceccherini V (1961b) Organizzazioni isotipiche nelle riaggregazioni in vitro di abbozzi disgregati di retina di embrioni di pollo. Rend Accad Naz Lincei 8 30:818–822

Stefanelli A, Zacchei AM, Ceccherini V (1961c) Ricostituzioni retiniche in vitro dopo disgregazione dell'abbozzo oculare di embrione di pollo. Acta Embryol Morphol Exp 4:47–55

Stefanelli A, Zacchei AM, Caravita S, Cataldi E, Ieradi LA (1966a) Sinapsi in vitro da cellule disgregate di retina di embrione di pollo. Rend Accad Naz Lincei 8 60:758–762

Stefanelli A, Zacchei AM, Caravita S, Cataldi E, Ieradi LA (1966b) Differenziamento in riaggregati di retina in vitro. Arch Ital Zool 51:985–996

Stefanelli A, Zacchei AM, Caravita S, Cataldi E, Ieradi LA (1967a) Differenziamento di fotorecettori di pollo in riaggregati coltivati in vitro. Rend Accad Naz Lincei 8 62:594–597

Stefanelli A, Zacchei AM, Caravita S, Cataldi E, Ieradi LA (1967b) New forming retinal synapses in vitro. Experientia 23:199–200

Stell WK (1972) The morphological organization of the vertebrate retina. In: Fuortes F (ed) Handbook of sensory physiology, vol VII/2, Physiology of photoreceptor organs. Springer, Berlin Heidelberg New York, pp 111–215

Straznicky K, Gaze RM (1971) The growth of the retina in Xenopus laevis: An autoradiographic study. J Embryol Exp Morphol 26:67–79

Sugiyama J, Daniels MP, Nirenberg M (1977) Muscarinic acetylcholine receptors of the developing retina. Proc Natl Acad Sci USA 74:5524–5528

Suzuki O, Noguchi E, Yagi K (1977) Monoamine oxidase in developing chick retina. Brain Res 135:305–314

Tamai M, Chader GJ (1979) Early appearance of disk shedding in the rat retina. Invest Ophthalmol Vis Sci 18:913–929

Tamai M, Takahashi J, Noji T, Mizuno K (1978) Development of photoreceptor cells in vitro: Influence and phagocytotic activity of homo- and heterogenic pigment epithelium. Exp Eye Res 26:581–590

Tansley K (1933) The formation of rosettes in the rat retina. Br J Ophthalmol 17:321–336

Tewari HB, Tyagi HR (1968) On the alkaline phosphatase activity amongst the constituents of the eye of Calotes versicolor in light and dark environments. Exp Eye Res 7:200–204

Tilney F, Casamajor L (1924) Myelinogeny as applied to the study of behavior. Arch Neurol Psychiatry 12:1–66

Tomita T (1978) ERG waves and retinal cell function. Sen Processes 2:276–284

Towlson MJ (1964) Calcium and the adenosine triphosphatase activity of the developing rat retina. Nature 201:295–296

Tucker GS, Hollyfield JG (1977) Modifications by light of synaptic density in the inner plexiform layer of the toad Xenopus laevis. Exp Neurol 55:133–151

Tucker GS, Hamasaki DI, Labbie A, Muroff J (1979) Anatomic and physiologic development of the photoreceptor of the kitten. Exp Brain Res 37:459–474

Tunnicliff G, Firneisz G, Ngo TT, Martin RO (1975) Developmental changes in the kinetics of γ-aminobutyric acid transport by chick retina. J Neurochem 25:649–652

Ueno K-J (1961) Morphogenesis of the retinal cone studied with the electron microscope. Jpn J Ophthalmol 5:114–122

Ulshafer RJ, Clavert A (1979) Modification de la fixation de la concanavalin A au cours du développement de la rétine chez Xenopus laevis. C R Soc Biol (Paris) 173:127–132

Ulshafer RJ, Clavert A (1980) In vitro studies on differentiation of the optic stalk in the chick embryo. Differentiation 16:189–192

Utermann D (1964) Biochemische Untersuchungen über die Entwicklung des Stoffwechsels in der Rattennetzhaut. Albrecht von Graefes Arch Ophthalmol 167:541–548

Vinnikov YA (1969) The ultrastructural and cytochemical basis of the mechanism of function of the sense organ receptors. In: Bourne GH (ed) The structure and function of nervous tissue, vol 2. Academic Press, New York London, pp 265–392

Vizi ES (1978) Na^+-, K^+-activated adenosine triphosphatase as a trigger in transmitter release. Neuroscience 3:367–385

Vogel M (1978a) Postnatal development of the cat's retina. Adv Anat Embryol 54:1–66

Vogel M (1978b) Postnatal development of the cat's retina – concept of maturation obtained by qualitative and quantitative examinations. Albrecht von Graefes Arch Ophthalmol 208:93–108

Vogel Z, Nirenberg M (1976) Localization of acetylcholine receptors during synaptogenesis in retina. Proc Natl Acad Sci USA 73:1806–1810

Vogel Z, Daniels MP, Nirenberg M (1976) Synapse and acetylcholine receptor synthesis by neurons dissociated from retina. Proc Natl Acad Sci USA 73:2370–2374

Wagner H-J (1972) Vergleichende Untersuchungen über das Muster der Sehzellen und Horizontalen in der Teleostier-Retina (Pisces). Z Morphol Tiere 72:77–131

Wagner H-J (1974) Development of the retina of Nannacara anomala (Regan) (Cichlidae, Teleostei) with special reference to regional variations of differentiation. Z Morphol Tiere 79:113–132

Wainwright SD (1979) Development of hydroxyindole-o-methyltransferase activity in the retina of the chick embryo and young chick. J Neurochem 32:1099–1101

Wang G-K, Schmidt J (1976) Receptors for α-bungarotoxin in the developing visual system of the chick. Brain Res 114:524–530

Weidman TA, Kuwabara T (1968) Postnatal development of the rat retina. An electron microscopic study. Arch Ophthalmol 79:470–484

Weidman TA, Kuwabara T (1969) Development of the rat retina. Invest Ophthalmol 8:60–69

Wender M (1972) Maturation of neurons in the light of cytoenzymatic studies. Int J Neurosci 4: 131–137

Weysse AW, Burgess WS (1906) Histogenesis of the retina. Am Nat 40:611–637

Wiggert BO, Chader GJ (1975) A receptor for retinol in the developing retina and pigment epithelium. Exp Eye Res 21:143–151

Wilson MA (1971) Optic fibre counts and retinal ganglion cell counts during development of Xenopus laevis (Daudin). Q J Exp Physiol 56:83–91

Wilt FH (1959) The differentiation of visual pigments in metamorphosing larvae of Rana catesbeiana. Dev Biol 1:199–233

Wise C, Kunz YW (1977) Ultrastructural morphogenesis of cones and rods in the teleost Poecilia reticulata P. In: 8th Int. Congr. Int. Soc. Develop. Biol. Tokyo 1977, Abstract C 0202

Witkovsky P (1963) An ontogenetic study of retinal function in the chick. Vision Res 3:341–355

Witkovsky P, Gallin E, Hollyfield JG, Ripps H, Bridges CDB (1976) Photoreceptor thresholds and visual pigment levels in normal and vitamin A-deprived Xenopus tadpoles. J Neurophysiol 39:1272–1287

Wolff H (1969) Über die Entwicklung des Enzymmusters der Rattenretina. Histochemie 17:11–29

Yacob A, Wise C, Kunz YW (1977) The accessory outer segment of rods and cones in the retina of the guppy, Poecilia reticulata P. (Teleostei). Cell Tissue Res 177:181–193

Yamada E, Ishikawa T (1965) Some observations on the submicroscopic morphogenesis of the human retina. In: Rohen J (ed) The structure of the eye. Schattauer, Stuttgart, pp 5–16

Yew DT (1976) The developmental histochemistry of the chicken visual cells. Acta Anat 96:561–567

Yew DT (1979) Autoradiographic study of neonatal rat retina after labelled uridine uptake. Acta Histochem 64:1–4

Yew DT, Chan Y (1977) The effect of laser on the developing rodent retinas. Acta Anat 99:386–390

Yew DT, Meyer DB (1975) Two types of bipolar cells in the chick retina development. Experientia 31:1077–1078

Yew DT, Ho AKS, Meyer DB (1974) Effect of 6-hydroxydopamine on retinal development in the chick. Experientia 30:1320–1322

Yew DT, Yoshihara HM, Meyer DB (1975) Localization of ATPase in the choroid and retina of the developing chick. Experientia 31:846–848

Yoshida M, Ninomiya N (1967) Electron microscopy of retinal rods in frog larvae with special reference to the oil droplet. Annot Zool Jpn 40:91–97

Young RW (1976) Visual cells and concept of renewal. Invest Ophthalmol 15:700–725

Zavarzin AA, Stroeva OG (1964) A study of DNA synthesis and kinetics of cell population at differentiation of the retina and pigment epithelium and the iris by the 3H-thymidine method. In: Academy of Sciences USSR (ed) Investigations on cell cycles and metabolism of nucleic acids during cell differentiation (in Russian). Moscow, pp 116–125

Zetterström B (1956) The effect of light on the appearance and development of the electroretinogram in newborn kitten. Acta Physiol Scand 35:272–279

Zimmermann LE, Eastham AB (1959) Acid mucopolysaccharide in the retinal pigment epithelium and visual cell of the developing mouse eye. Am J Ophthalmol 47:488–499

Subject Index

α-bungarotoxin receptors 14f, 49f, 66
Accessory outer segment 40
Acetylcholine, ACh 8, 14f, 49f, 66
Acetylcholinesterase, AChE 14f, 20, 24, 46, 49f, 62f, 66
Acetylcholine receptors 14, 49f, 59
Adenylate cyclase 20, 24, 52
Adenosine triphosphatase ATPase 5, 15, 20, 24, 52, 62f
Astaxanthine 16, 38, 64
a-wave 8, 16, 20, 24, 29, 55f, 66

b-wave 8, 16f, 20, 24, 29, 55f, 66

Calycal processes 13, 36
Catecholamines 5, 12, 50, 62
Centriole 10, 13, 21, 28, 36, 40, 59
Ciliary connection, stalk, outgrowth 5, 7, 13, 19, 38, 40, 64
Cilium 13, 28, 40, 59
Chievitz, layer of 26, 46
c-wave 16f, 55f, 66
Cyclic AMP 15, 20, 51
Cyclic GMP 15, 51

Darkness 41, 48, 57
Desmosomes 10, 13, 18, 28, 65
Disks 5, 8, 13, 19, 23, 32, 38ff, 57, 59, 64
Diaphorase 18, 22, 62ff
DNA 6, 9
Double cones 5, 10, 11, 38, 40, 59, 64
d-wave 16, 55f
Dyad 19, 43, 45

Ellipsoid 5, 7, 13, 19, 22, 23, 29, 32, 36
Endoplasmic reticulum 3, 5, 6, 10, 13, 21, 23, 28f, 36f, 43, 59, 62ff

Galloxanthine 38, 64
γ-amino butyric acid, GABA 8, 15, 20, 24, 31, 34, 50f, 66
Gangliosids 16, 51, 66
Gap junctions 6, 10, 11, 64f
Glucosamine 7, 65
Glutamate 24, 50f
Glutamine 15, 20, 24, 66

Glycine 20, 24, 50f, 66
Glycogen 13, 36f, 64
Golgi apparatus 3, 5, 6, 13, 19, 21, 23, 28, 36, 59, 62ff
Growth cones 5, 21

Hydroxyindole-o-methyltransferase, HIOMT 24, 50

Indoleamines 15, 50

Light 41, 48, 52, 57

Macula adhaerentia 6, 11, 65
Microtubuli 6, 7, 13, 21, 28, 36, 43
Microvilli 10, 13, 36, 64
Mitochondria 3, 5, 10, 13, 21, 23, 28f, 36, 59, 62ff
Mitoses 3, 6, 9, 17, 20, 26, 29, 65
Monoaminoxidase, MAO 15, 50, 66
Myelination 7, 11, 28, 54f, 62
Myoid 7, 13, 14, 53

Neurofilaments 5, 21, 46, 62f
Neurotubuli 5, 46, 62
Nissl substance 6, 28, 6sf

Oil droplet 7, 13, 16, 38, 59, 64
Opsin 41
Optic cup 9, 17, 18, 20

Paraboloid 13, 36, 59, 64
PAS-positive substances 13, 18, 64
Phagosomes 5, 8, 23, 41, 64, 66
Phosphatases, acid, alkaline 3, 11, 21, 62f
Phospholipids 15, 51, 66
Porphyropsin 8, 41
Punctate junctions 10, 11, 13, 65

Receptive fields 9, 53f, 62
Responses to light 9, 16, 25, 29, 32, 54, 66
Retinol 16, 40, 59

84

Advances in Anatomy, Embryology and Cell Biology

Editors:
F. Beck, W. Hild,
J. van Limborgh, R. Ortmann,
J. E. Pauly, T. H. Schiebler

Springer-Verlag
Berlin
Heidelberg
New York

A. Raedler, J. Sievers

The Development
of the Visual System
of the Albino Rat

1975. 16 figures. 88 pages (Advances in Anatomy, Embryology and
Cell Biology, Volume 50, Part 3). ISBN 3-540-07079-6

Light- and electromicroscopic studies were made of the induction of
the development of the rat visual systems with special reference to the
correlation between cytological and histological structure, also experimental studies with the brain edema-producing antimetabolite, 6-
aminonicotineamide. The effects of the drug on the undifferentiated
ventricular cells and the differentiating glia and nerve cells are
discussed together with the consequences they have on the development of the visual system.

H. Wolburg

Axonal Transport,
Degeneration,
and Regeneration in the
Visual System of the Goldfish

1981. 28 figures. IX, 94 pages (Advances in Anatomy, Embryology
and Cell Biology, Volume 67). ISBN 3-540-10336-8

Slow and fast axonal transport, neuronal and Wallerian degenaration,
neuronal and axonal regeneration and dependencies both between
perikaryon and axon and neuron and glial cell are described. Axoglial
substance transfer and the myelinating activity of the oligodendroglia
in the regenerating system as reaction to neuronal conditions are
demonstrated and discussed within the framework of current views.

M. Vogel

Postnatal Development
of the Cat's Retina

1978. 27 figures, 2 tables. 66 pages (Advances in Anatomy, Embryology and Cell Biology, Volume 54, Part 4). ISBN 3-540-08799-0

The purpose of this volume is to present coherently, largely on
electron microscopic basis, the various differentiation processes of
retinal layers and individual structures taking place in the cat's retina
during the postnatal period. These developmental processes are
examined both in a qualitative-morphological and in a quantitative-
morphometrical context. The data obtained by this study exhibit
various differentiation phases and alternating growth trends between
the different retinal structures and layers in summary, a centripetal
maturation of the cat's retina. Special attention is given qualitatively to
the photo-lamelae genesis, the relationship between pigment epithelium and receptors and the morphogenetic cellular decay.

Springer-Verlag
Berlin
Heidelberg
New York